Collins

INTERNATIONAL PRIMARY MATHS

Workbook 4

William Collins' dream of knowledge for all began with the publication of his first book in 1819. A self-educated mill worker, he not only enriched millions of lives, but also founded a flourishing publishing house. Today, staying true to this spirit, Collins books are packed with inspiration, innovation and practical expertise. They place you at the centre of a world of possibility and give you exactly what you need to explore it.

Collins. Freedom to teach.

Published by Collins
An imprint of HarperCollins*Publishers*
The News Building
1 London Bridge Street
London
SE1 9GF

HarperCollins*Publishers*
Macken House,
39/40 Mayor Street Upper,
Dublin 1,
D01 C9W8, Ireland

Browse the complete Collins catalogue at
www.collins.co.uk

© HarperCollins*Publishers* Limited 2021

10 9 8 7 6

ISBN 978-0-00-836948-4

British Library Cataloguing-in-Publication Data
A catalogue record for this publication is available from the British Library.

Author: Caroline Clissold
Series editor: Peter Clarke
Publisher: Elaine Higgleton
Product developer: Holly Woolnough
Project manager: Mike Harman (Life Lines Editorial Services)
Development editor: Joan Miller
Copyeditor: Tanya Solomons
Proofreader: Tanya Solomons
Cover designer: Gordon MacGilp
Cover illustrator: Ann Paganuzzi
Typesetter: Ken Vail Graphic Design LTD
Illustrators: Ann Paganuzzi, Ken Vail Graphic Design and QBS Learning
Production controller: Lyndsey Rogers
Printed and bound in India by Replika Press Pvt. Ltd.

With thanks to the following teachers and schools for reviewing materials in development: Antara Banerjee, Calcutta International School; Hawar International School; Melissa Brobst, International School of Budapest; Rafaella Alexandrou, Pascal Primary Lefkosia; Maria Biglikoudi, Georgia Keravnou, Sotiria Leonidou and Niki Tzorzis, Pascal Primary School Lemessos; Taman Rama Intercultural School, Bali.

MIX
Paper | Supporting responsible forestry
FSC™ C007454

This book is produced from independently certified FSC™ paper to ensure responsible forest management.

For more information visit: **www.harpercollins.co.uk/green**

The publishers gratefully acknowledge the permission granted to reproduce the copyright material in this book. Every effort has been made to trace copyright holders and to obtain their permission for the use of copyright material. The publishers will gladly receive any information enabling them to rectify any error or omission at the first opportunity.

Contents

Number

Geometry and Measure

Statistics and Probability

How to use this book

This book is used during the middle part of a lesson when it is time for you to practise the mathematical ideas you have just been taught.

- An **objective** explains what you should know, or be able to do, by the end of the lesson.

You will need
- Lists the resources you need to use to answer some of the questions.

There are two pages of practice questions for each lesson, with three different types of question:

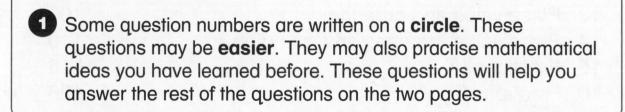 Some question numbers are written on a **circle**. These questions may be **easier**. They may also practise mathematical ideas you have learned before. These questions will help you answer the rest of the questions on the two pages.

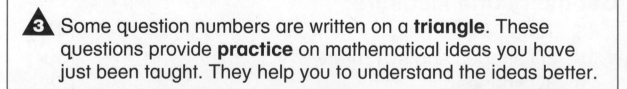 Some question numbers are written on a **triangle**. These questions provide **practice** on mathematical ideas you have just been taught. They help you to understand the ideas better.

5 Some question numbers are written on a **square**. These questions are slightly more **challenging**. They make you think more deeply about the mathematical ideas.

You won't always have to answer all the questions on the two pages. Your teacher will tell you which questions to answer.

HINT
Draw a ring around the question numbers your teacher tells you to answer.

 Questions with a star beside them require you to Think and Work Mathematically (TWM). You might want to use the TWM Star at the back of the Student's Book to help you.

Date: _____

At the bottom of the second page there is room to write the date you completed the work on these pages. If it took you longer than 1 day, write all of the dates you worked on these pages.

Self-assessment

Once you've answered the questions on the pages, think carefully about how easy or hard you find the ideas. Draw a ring around the face that describes you best.

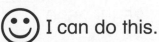 I can do this.

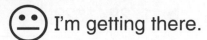 I'm getting there.

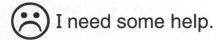 I need some help.

Number

Lesson 1: **Counting in 10s, 100s and 1000s**

• Count on and back in steps of 10, 100 and 1000

1 Count on in 100s from 167.

167, [   ], [   ], [   ], [   ], [   ], 767

2 Count back in 100s from 945.

945, [   ], [   ], [   ], [   ], [   ], 345

3 Write the missing numbers as you count on or back in 1000s.

a 1597, 2597, [   ], [   ], [   ], [   ], 7597

b 9305, 8305, [   ], [   ], [   ], [   ], [   ]

c 7981, 8981, [   ], [   ], [   ], 12 981

4 Write the missing numbers as you count on or back in 100s.

a 2590, 2690, [   ], [   ], [   ], 3090, [   ]

b 1729, 1629, [   ], 1429, [   ], [   ], [   ]

c 10 312, 10 212, [   ], [   ], [   ], [   ]

5 Write the missing numbers as you count on or back in 10s.

a 4663, 4673, [   ], [   ], [   ], [   ], 4723

b 9641, 9631, 9621, [   ], [   ], [   ], [   ]

c 7180, [   ], [   ], [   ], [   ], 7230, 7240

Number

6 Adah counts back in 1000s for six counts. She lands on 1675.

What number does she start from?

_____

How do you know?

_____

_____

7 Samson counts on in 100s from 934.

a What is the sixth number he says? [   ]

b What is the tenth number he says? [   ]

8 Write the numbers in the boxes as you count back in 100s.

1245, [   ], [   ] 945, [   ], [   ] 645, [   ],

[   ], [   ], [   ], [   ], [   ]

9 Jayden counts back to 24 in hundreds. He says 12 numbers.

What is his starting point? [   ]

Explain how you know.

_____

_____

10 Serina thinks that if she counts in hundreds, she will say 200.

Is this always, sometimes or never true? [   ]

Explain how you know.

_____

_____

Date: _____   ☺ 😐 ☹

Lesson 2: **Adding odd and even numbers**

• Recognise and explain rules for adding odd and even numbers

1 Colour the odd numbers yellow and the even numbers green.

1	2	3	4	5	6	7	8	9	10
11	12	13	14	15	16	17	18	19	20
21	22	23	24	25	26	27	28	29	30
31	32	33	34	35	36	37	38	39	40
41	42	43	44	45	46	47	48	49	50

You will need
• yellow and green pencils

2 What do you notice about your grid in **1**?

_____

_____

3 Choose pairs of numbers from the Carroll diagram.

Write whether they will give an even or odd sum.

Even numbers		Not even numbers	
246	370	327	571
528	854	679	743

Don't work out the sums!

a [ ] + [ ] = [ ]

b [ ] + [ ] = [ ]

c [ ] + [ ] = [ ]

d [ ] + [ ] = [ ]

4 **a** Jafan wants to know how to spot odd and even numbers.

What are the rules for identifying odd and even numbers?

_____

Number

b Rhian says:

> Whether a number is odd or even has something to do with the 2 times table.

Is she right? ☐

Explain your answer.

_____

_____

 5 Complete each calculation and write **E** or **O** underneath each number.

Example: 10 + 12 = 22

☐ E ☐ + ☐ E ☐ = ☐ E ☐

a 17 + 12 = ☐

☐ + ☐ = ☐

b 24 + 16 = ☐

☐ + ☐ = ☐

c 12 + 17 = ☐

☐ + ☐ = ☐

d 23 + 25 = ☐

☐ + ☐ = ☐

6 Tick the statements that you think are true.

a An even number plus an even number gives an odd number. ☐

b Two odd numbers added together give an even number. ☐

c The total of an odd number and an even number depends on the order you add them together. ☐

7 Mara thinks that if she adds 257 and 193 the sum will be odd as there are more odd digits than even.

Is she correct? ☐ Explain your thinking.

_____

_____

Date: _____

Number

Lesson 3: **Subtracting odd and even numbers**

• Recognise and explain rules for subtracting odd and even numbers

1 Complete each calculation and write **E** or **O** underneath each number.

Example: 23   –   10   =   13

$\boxed{O}$ – $\boxed{E}$ = $\boxed{O}$

a   12   –   6   = $\boxed{}$

$\boxed{}$ – $\boxed{}$ = $\boxed{}$

b   24   –   13   = $\boxed{}$

$\boxed{}$ – $\boxed{}$ = $\boxed{}$

c   11   –   9   = $\boxed{}$

$\boxed{}$ – $\boxed{}$ = $\boxed{}$

d   18   –   14   = $\boxed{}$

$\boxed{}$ – $\boxed{}$ = $\boxed{}$

e   13   –   7   = $\boxed{}$

$\boxed{}$ – $\boxed{}$ = $\boxed{}$

f   17   –   12   = $\boxed{}$

$\boxed{}$ – $\boxed{}$ = $\boxed{}$

2 Choose pairs of numbers from the Carroll diagram.

Will they give an even or odd difference?

Even numbers		Not even numbers	
2466	3702	1327	1571
4528	7854	1679	1743

Don't work out the differences!

a $\boxed{}$ – $\boxed{}$ = $\boxed{}$

b $\boxed{}$ – $\boxed{}$ = $\boxed{}$

c $\boxed{}$ – $\boxed{}$ = $\boxed{}$

d $\boxed{}$ – $\boxed{}$ = $\boxed{}$

Number

3 Below each addition and subtraction there are two answers.
Circle the correct answer.

a 568 + 227 =

| 795 | 794 |

b 825 − 543 =

| 282 | 283 |

c 553 + 435 =

| 987 | 988 |

d 854 − 327 =

| 527 | 528 |

e 236 + 452 =

| 689 | 688 |

f 258 − 126 =

| 133 | 132 |

g 747 − 319 =

| 428 | 427 |

h 543 + 216 =

| 758 | 759 |

i 881 − 503 =

| 378 | 377 |

4 Holly says:

> I find it really useful to know what happens when I add or subtract different combinations of even and odd numbers.

Explain why Holly thinks this.

_____

_____

 5 Carla says:

> When you subtract numbers, the difference between them is always going to be even.

a Subtract different combinations of even and odd numbers to test Carla's thinking. Write them in the box below.

b Is Carla correct? [   ] Explain your answer.

_____

_____

Date: _____

Number

Lesson 4: **Sequences**

• Recognise and extend number sequences

1 Draw lines to match each number sequence to the rule that describes it.

1, 3, 5, 7, 9, 11	• Increases by 3
5, 10, 15, 20, 25, 30	• − 2
60, 50, 40, 30, 20, 10	• + 2
5, 8, 11, 14, 17, 20	• − 4
22, 18, 14, 10, 6, 2	• + 5
28, 26, 24, 22, 20, 18	• Decreases by 10

2 Make up your own number sequence where the numbers = decrease by 3. Include 10 numbers in your sequence.

_____

_____

3 These number sequences are based on the multiples of different numbers. Which times table does each sequence show?

a 6, 12, 18, 24, 30   These are all multiples of ☐.

The rule is [ ].

b 40, 36, 32, 28, 24   These are all multiples of ☐.

The rule is [ ].

c 35, 30, 25, 20, 15   These are all multiples of ☐.

The rule is [ ].

d 54, 45, 36, 27, 18   These are all multiples of ☐.

The rule is [ ].

Number

4 Write the next two numbers in each sequence. Then write the rule.

a 30, 25, 20, 15, 10, ☐, ☐

The rule is _____.

b 53, 60, 67, 74, 81, ☐, ☐

The rule is _____.

c 22, 33, 44, 55, 66, ☐, ☐

The rule is _____.

d 45, 41, 37, 33, 29, ☐, ☐

The rule is _____.

5 Sacha makes up a number sequence. It begins on the number 10. It ends on the number 50. There are 9 numbers in his number sequence.

a What is his sequence?

_____

b Describe Sacha's rule.

_____

6 Beatrice makes up a number sequence. It begins on 42. It ends on –12. There are 10 numbers in her number sequence.

a What is her sequence?

_____

b Describe Beatrice's rule.

_____

Date: _____

Lesson 1: **Counting in 1-digit steps**

Number

• Count on and back in 1-digit steps of constant size

1 **a** Count on in steps of 4 from 0 to 40.

0, ☐, ☐, ☐, ☐, ☐, ☐, ☐, ☐, ☐, 40

b Count back on in steps of 8 from 80 to 0.

80, ☐, ☐, ☐, ☐, ☐, ☐, ☐, ☐, ☐, 0

c Which numbers appear in both the counts of 4 and 8?

_____

2 **a** Count on in steps of 9 from 0 to 90.

0, ☐, ☐, ☐, ☐, ☐, ☐, ☐, ☐, ☐, 90

b Use the pattern you can see to count on in 9s from 1 to 91.

1, ☐, ☐, ☐, ☐, ☐, ☐, ☐, ☐, ☐, 91

c Use the pattern to count on in 9s from 5 to 95.

5, ☐, ☐, ☐, ☐, ☐, ☐, ☐, ☐, ☐, 95

d Describe the pattern.

_____

_____

3 Petra says: If I count on in 6s from zero, every number will be even.

Do you agree? ☐ Why? _____

_____

4 Mia says: If you add together the digits of each product in the 9 times table they have a total of 9. That means if I count in 9s from zero, I know I will say 72 because 7 add 2 equals 9.

Tibs says: That means that if you count in 9s from zero, you will say 135.

Do you agree? ☐ Why?

_____

_____

Prove it! _____

_____

_____

5 a Why do the numbers alternate between odd and even when you count in steps of 9?

_____

_____

b Why are the numbers all even when you count in steps of 8?

_____

_____

c Is it possible to have a count of only odd numbers when you count in steps from zero? ☐ Explain.

_____

_____

Date: _____

Number

Lesson 2: **Counting in 1-digit steps beyond zero**

> • Count on and back in 1-digit steps beyond zero

1 Write the numbers in the correct positions on the number line.

| −5 | 7 | −1 | 3 | −8 | 4 |

−10   0   10

2 Tia counts on in steps of 4 from 0. This is what she says:

0, 4, 8, 12, 16, 20, 28, 32, 36, 40

She has missed out a number.

Which number did she miss?

3 Amare counts on in steps of 6 from 0. These are the numbers that he says:

0, 6, 12, 18, 24, 30, 36, 42, 48, 54, 60

He then counts back in steps of 6 from 0.

After saying 0, what are the next ten numbers he says?

4 Sanjit thinks that if he counts back in steps of 9 from 4, the first number that he will say must be even.

Do you agree?

Use your knowledge of adding and subtracting odd and even numbers to explain why.

5 Suzie says: I counted back in steps of 8 from 32 for 9 steps and I landed on –40.

Kwame says: If I count back in steps of 10 from 40, I will also land on –40. It will take me fewer steps.

Do you agree? ☐ Explain.

_____

_____

6 Tess says: If I count back in 5s from 20 to –30, I will always say odd numbers.

Is what Tess says always, sometimes or never true? ☐
Explain.

_____

_____

7 Heema thinks that if she counts back from 72 in steps of 9 for 10 steps, she will land on an even number.

Is she correct?

Explain your thinking.

_____

_____

What number will she land on? ☐

8 Jared thinks that 1000 is exactly the same number of steps away from zero as –1000.

Is Jared correct? ☐ Explain.

_____

_____

Date: _____   ☺ 😐 ☹

Number

Lesson 3: **Sequences**

• Recognise and extend number sequences

1 The rule for this sequence is add 5 to the previous number.
Write the next four terms.

20,   25,   30,   35,   ☐,   ☐,   ☐,   ☐

2 The rule for this sequence is subtract 3.
Write the next four terms.

6,   3,   0,   ☐,   ☐,   ☐,   ☐

3 What is the rule for each of these sequences?

a 24,   12,   6,   3

_____

b 162,   54,   18,   6,   2

_____

c 100,   50,   0,   −50,   −100,   −150

_____

4 Complete this number sequence.

144,   ☐,   36,   ☐,   9,   ☐

What is the rule?

_____

5 Make up your own number sequence.

☐

Explain the rule.

_____

Number

6 Bao-Yu makes up a sequence. It has eight numbers. It begins on 100 and finishes on –40.

Write her sequence.

What is her rule?

_____

7 Carla says:

> This is my number sequence.
> I made it by halving.

40,   20,   10,   5,   0,   –5,   –10,   –20,   –40

Carla's sequence is incorrect.

a What has she done wrong?

_____

_____

b What should the number after 5 be if her rule is halving the previous term? ☐ Why?

_____

_____

8 Mercy wrote this sequence: 2, 4, 5, 10, 11, 22, 23, 46.

Her teacher gave her a rule to follow.

a What was that rule?

_____

_____

b Write the next five numbers in Mercy's sequence.

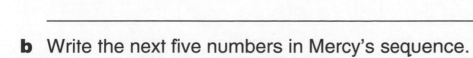 , , , ,

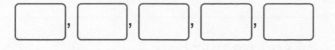

Date: _____

Lesson 4: **Square numbers**

- Recognise and extend the pattern of square numbers

1 Draw a ring around the square numbers.

   1   4   10   25   17   9

2 Draw lines to match each square number to the correct times table.

9	6 × 6
16	2 × 2
4	5 × 5
1	3 × 3
25	1 × 1
36	4 × 4

3 Draw diagrams to show these square numbers.

Label each diagram with the times table fact that shows the square number.

4   9   16

4 Complete this sequence of square numbers.

   4, ☐, 16, ☐, 36, 49, ☐, 81, ☐

Number

5 Sadie says:

I think that every time we multiply two numbers together, we will get a square number.

Do you agree? [   ] Explain your thinking.

_____

_____

6 Draw the next two terms in this pattern.

[grid with pattern]

[   ]   [   ]   [   ]   [   ]

Label each part of the sequence with the times table fact.

7 Faith draws a diagram to show a square number.

She thinks that because a square is a rectangle, this rectangle must also show a square number.

Explain to Faith what a square number is and why she is incorrect.

_____

_____

_____

_____

Date: _____

Number

Lesson 1: **Reading and writing numbers to 1000**

* Read and write numbers to 1000

1 Write these numbers in numerals.

a one hundred and sixty-five [   ]

b one hundred and thirty-seven [   ]

c one hundred and five [   ]

2 Draw lines to match each numeral to a written number.

189   one hundred and thirty

148   one hundred and eighty-nine

176   one hundred and seventy-six

109   one hundred and forty-eight

130   one hundred and nine

What do all these numbers have in common?

_____

3 Put the words together to make two different 3-digit numbers.

a hundred   sixty   three   and   two

_____

_____

b forty   hundred   and   eight   one

_____

_____

c five   nine   hundred   thirty   and

_____

_____

Number

4 Natalia has these digit cards: | 6 | 7 | 8 |

She says:

> It is impossible to make a number below 600 with my cards.

Is she correct? ☐ Explain.

_____

_____

5 Write these numbers in words.

a 356 _____

b 437 _____

c 898 _____

d 906 _____

6 Use the digit cards to make as many different 3-digit numbers as you can.

| 7 | 5 | 2 | 1 | Write the numbers in numerals.

7 Feechi is thinking of a number. Each digit is an even number. All the digits are different. His number is greater than 800. What number could he be thinking of? Write all the possibilities.

Lesson 2: **Reading and writing numbers to 10 000**

• Read and write numbers to 10 000

1 Draw lines to match each number with the correct number of thousands.

7189	five thousands
5462	seven thousands
1783	one thousand

2 Draw lines to match each number with the correct number of hundreds.

5677	eight hundreds
1046	six hundreds
7832	zero hundreds

3 Complete the numbers.

a 3254: _____ thousand, _____ hundred and _____ four

b 6707: six _____, seven _____ _____ seven

c 4198: four _____, one _____ _____ ninety _____

d 9387: _____ thousand, _____ hundred and _____ _____

e 1060: _____ _____ and _____

4 Alfie is thinking of a 4-digit number. All the digits are different. It is less than 2000. The hundreds are less than 2. The tens are greater than 8. The ones are between 2 and 4.

a Write your answer in numerals. ☐

b Write the number in words.

5 Identify these numbers. Write them in numerals and in words.

a 2 ones, 8 thousands, 6 hundreds, 5 tens

[ ] _____

_____

b 9 hundreds, 7 ones, no tens, 3 thousands

[ ] _____

_____

c 4 ones, no hundreds, 3 thousands, 8 tens

[ ] _____

_____

d no tens, no hundreds, 5 thousands, 3 ones

[ ] _____

_____

6 Samson is thinking of a number. The digits in his number are 2, 4, 6 and 7. His number is an odd number. What number might he be thinking of? Write the numbers in numerals and in words.

How many possibilities are there? [ ]

Date: _____

Number

Lesson 3: **Reading and writing numbers to 100 000**

• Read and write numbers to 100 000

1 Underline the thousands. Draw a ring around the tens.

twenty-one thousand, three hundred and forty-four

twenty-two thousand, five hundred and thirty-seven

twenty-three thousand, four hundred and fifty-nine

twenty-eight thousand, one hundred and seventy-one

2 Write these numbers in words.

a 11 243 _____

b 11 244 _____

c 11 245 _____

3 Mia writes: fifty-six thousand, three hundred and twenty-one

She says that she can use the same digits to make another number that is greater than this one.

Do you agree? ☐ Why? Show some examples.

_____

_____

4 Write these numbers in numerals.

a forty-two thousand, three hundred and sixteen ☐

b seventy-two thousand, five hundred and eight ☐

c fourteen thousand, six hundred and forty-two ☐

d sixty-six thousand and eighteen ☐

e seventy-five thousand and two ☐

5 Ekon uses this grid to make a number. He colours the numbers he uses. What number has he made?

1	2	3	4	5	6	7	8	9
10	20	30	40	50	60	70	80	90
100	200	300	400	500	600	700	800	900
1000	2000	3000	4000	5000	6000	7000	8000	9000
10 000	20 000	30 000	40 000	50 000	60 000	70 000	80 000	90 000

a Write the number in numerals.

b Write it in words. _____

c Use the grid to make up three different numbers of your own. Write each number in numerals and words.

6 Here are Ben's vocabulary cards. He uses all these cards to make up four different numbers.

fourteen	ninety	thousand	hundred

and	three	eight

a Write the four numbers in numerals and words.

b Why can't he make any more numbers?

_____

_____

_____

Date: _____

Number

27

Number

Lesson 4: **Negative numbers**

- Read, write and count with negative numbers

1 Write the temperature shown on each thermometer.

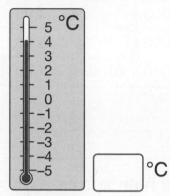 **a** [ ]°C

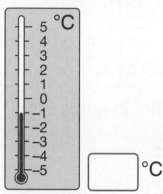 **b** [ ]°C

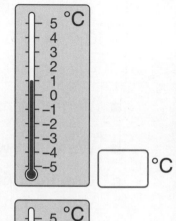 **c** [ ]°C

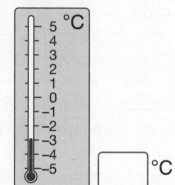 **d** [ ]°C

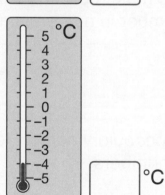 **e** [ ]°C

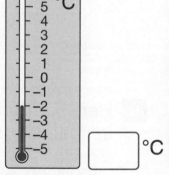 **f** [ ]°C

2 Order the temperatures on the thermometers in **1** from lowest to highest.

_____

3 Draw a line from each number to show where it belongs on the number line.

a   7   −1   −3   −8

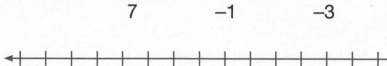

−10   0   10

b   −9   4   −4   −6

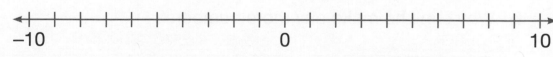

−10   0   10

Number

4 **a** The temperature is 2°C. Overnight it falls by 5 degrees.

What temperature is it now? ☐ °C

b The temperature is −7°C. It rises by 1 degree.

What temperature is it now? ☐ °C

c The temperature is 8°C during the day and −1°C during the night. By how many degrees does the

temperature fall? ☐ degrees

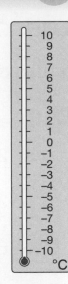

5 The floors of a building are labelled from 5 to −3. Floor 5 is the top floor. Floor 0 is the ground floor. Floors −1, −2 and −3 are all below ground level.

a You are on Floor 3 and travel down four floors.

What floor are you now on? ☐

b You enter the building at ground level and go down two floors. Is the floor you are on a

positive or a negative number? _____

c You are on Floor −2 and travel up three floors.

What floor are you now on? ☐

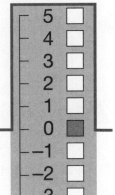

6 Sophie is thinking of a negative number. Petra is thinking of a positive number. If Sophie counts on 5 steps from her number, she will land on Petra's number. The girls' two numbers are between −10 and 10. What numbers could they be thinking of? Write all the possible numbers.

_____

_____

_____

_____

Date: _____

29

Lesson 1: **Mental addition**

Number

> • Use mental calculation strategies to add

1 Use a mental strategy to add these numbers. Estimate first.

a $25 + 26 =$ ☐ My estimate is ☐

b $36 + 19 =$ ☐ My estimate is ☐

c $68 + 21 =$ ☐ My estimate is ☐

2 Brady solves $145 + 99 =$. He starts on 145 and counts on 1 ninety-nine times.

Do you think his method is efficient? ☐ What should he do?

☐

3 Use a mental strategy to add these numbers.
Explain your strategies.

a $134 + 99 =$ ☐

☐

b $234 + 66 =$ ☐

☐

c $465 + 97 =$ ☐

☐

4 Find the unknown number.

a  $+ 200 = 656$  $=$ ☐   **b** $245 +$ 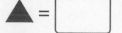 $= 745$ 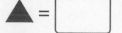 $=$ ☐

c $837 =$  $+ 537$  $=$ ☐   **d** $998 = 398 +$   $=$ ☐

Number

 Use a mental strategy to add these numbers.
Explain your strategies. Estimate first.

a 572 + 128 = [ ]   My estimate is [ ]

[ ]

b 564 + 111 = [ ]   My estimate is [ ]

[ ]

c 527 + 203 = [ ]   My estimate is [ ]

[ ]

6 Aki and Abdul discuss how to add 99 to another number.

Aki says: I will count on 99 to get my answer.

Abdul says: I will add 100 and subtract 1.

Who do you agree with? [ ] Why?

_____

_____

7 The numbers in each pair are the augend and sum of a calculation.

Which is the odd one out? [ ]

a 357 and 556   **b** 642 and 841

c 756 and 957   **d** 276 and 475

Why is this pair the odd one out?

_____

_____

Date: _____

Number

Lesson 2: **Mental addition involving money**

• Use mental calculation strategies to add money

1 Write the missing amounts

a ⬤ $+ 36c = 60c$ ⬤ $= \boxed{}$ b $59c +$ ▲ $= 80c$ ▲ $= \boxed{}$

c $97c = 50c +$ ■ ■ $= \boxed{}$ d $80c =$ ⬤ $+ 62c$ ⬤ $= \boxed{}$

2 Flo adds $99 to $234. She uses a written method.

Do you think that is sensible? $\boxed{}$ What could she do?

$\boxed{}$

3 Use a mental strategy to add these amounts.
Explain your strategies. Estimate first.

a $\$345 + \$98 = \boxed{}$ My estimate is $\boxed{}$

$\boxed{}$

b $\$186 + \$114 = \boxed{}$ My estimate is $\boxed{}$

$\boxed{}$

c $\$436 + \$142 = \boxed{}$ My estimate is $\boxed{}$

$\boxed{}$

d $\$299 + \$78 = \boxed{}$ My estimate is $\boxed{}$

$\boxed{}$

4 Find the missing amounts.

a ▲ $+ \$254 = \$454$ ▲ $= \boxed{}$ b $\$189 +$ ⬤ $= \$589$ ⬤ $= \boxed{}$

c $\$987 =$ ▲ $+ \$387$ ▲ $= \boxed{}$ d $\$865 = \$762 +$ ⬤ ⬤ $= \boxed{}$

Number

5 Use a mental strategy to add these amounts.
Explain your strategies. Estimate first.

a $326 + $299 = ⬚   My estimate is ⬚

⬚

b $462 + $138 = ⬚   My estimate is ⬚

⬚

c $453 + $212 = ⬚   My estimate is ⬚

⬚

d $267 + $103 = ⬚   My estimate is ⬚

⬚

6 Holly and Luke discuss how to add $86 to $214.

Holly says: I will count on $86 to get my answer.

Luke says: I will use my number bonds:
$6 + $4 = $10, $10 + $10 + $80 = $100.

Who do you agree with? ⬚ Why?

_____

_____

7 The amounts in each pair are the augend and total of a calculation.
Find the missing numbers and write the calculations.

a $326 and $546 ⬚

b $352 and $654 ⬚

Date: _____

Number

Lesson 3: **Mental subtraction**

• Use mental calculation strategies to subtract

1 Write the unknown number in each calculation.

   a ● – 19 = 40   ● = [   ]   **b** 63 – ▲ = 21   ▲ = [   ]

   c 25 = 80 – ■   ■ = [   ]   **d** 84 = ● – 16   ● = [   ]

2 Sunita subtracts 59 from 64. She starts at 64 and counts back in ones 59 times.

   Is this a good method? [   ]   What should she do?

   [   ]

3 Use a mental strategy to subtract these numbers.
   Explain your strategies. Estimate first.

   a 546 – 99 = [   ]   My estimate is [   ]

   [   ]

   b 764 – 52 = [   ]   My estimate is [   ]

   [   ]

   c 573 – 19 = [   ]   My estimate is [   ]

   [   ]

   d 382 – 79 = [   ]   My estimate is [   ]

   [   ]

4 Write the missing numbers.

   a ▲ – 254 = 454   ▲ = [   ]   **b** 985 – ● = 285   ● = [   ]

   c 73 = 473 – ■   ■ = [   ]   **d** 672 = ▲ – 200   ▲ = [   ]

Number

5 Use a mental strategy to subtract these numbers.
Explain your strategies. Estimate first.

a 754 – 399 = [ ]   My estimate is [ ]

[ ]

b 568 – 497 = [ ]   My estimate is [ ]

[ ]

c 835 – 623 = [ ]   My estimate is [ ]

[ ]

d 463 – 313 = [ ]   My estimate is [ ]

[ ]

6 Jeremiah and Faith discuss how to subtract 198 from 234.

Jeremiah says: I will subtract 100, then 90 and then 8 to get my answer.

Faith says: I will count on from 198 to 200 and then 234.

Who do you agree with? [ ] Why?

_____

_____

7 The numbers in each pair are the minuend and difference of a calculation. Find the subtrahend and write out the calculation.

a 654 and 152 [ ]

b 723 and 412 [ ]

Date: _____

35

Lesson 4: **Mental subtraction involving money**

Number

> • Use mental calculation strategies to subtract money

1 Subtract these amounts. Estimate first.

a 72c – 36c = ☐ My estimate is ☐

b 84c – 78c = ☐ My estimate is ☐

c 63c – 34c = ☐ My estimate is ☐

2 Ruby subtracts $99 from $121. She first subtracts $90 and then another $9. What else could she do that might be simpler?

☐

3 Use a mental strategy to subtract these amounts. Explain your strategies. Estimate first.

a $271 – $97 = ☐ My estimate is ☐

☐

b $345 – $21 = ☐ My estimate is ☐

☐

c $436 – $29 = ☐ My estimate is ☐

☐

d $786 – $79 = ☐ My estimate is ☐

☐

4 Write in the missing numbers.

a ✚ – $368 = $100 ✚ = ☐   **b** ■ – $763 = $200 ■ = ☐

c $795 – ▲ = $300 ▲ = ☐   **d** $694 – ● = $194 ● = ☐

Number

 5 Use a mental strategy to subtract these amounts.
Explain your strategies. Estimate first.

   a $782 – $131 = ⬚   My estimate is ⬚

   [   ]

   b $682 – $598 = ⬚   My estimate is ⬚

   [   ]

   c $872 – $603 = ⬚   My estimate is ⬚

   [   ]

6 Arjin and Eloise discuss how to subtract $297 from $459.

Eloise says: I will round $297 to $300, subtract that and then add $3 to get my answer.

Arjin says: I will use partitioning: $459 – $200 – $90 – $7.

Who do you agree with? ⬚ Why?

_____

_____

7 Work out the unknown amounts.

   a $168 – ⬚ = $92   **b** $768 – ⬚ = $392

   c ⬚ – $168 = $450   **d** ⬚ – $235 = $269

   e Find three possible values of ⬚ and ◯.

     Rules: Your values must be a 3-digit number and a 2-digit number. When they are subtracted there must be an exchange.

     ⬚ – ◯ = $238   ⬚ – ◯ = $238   ⬚ – ◯ = $238

Date: _____

Number

Lesson 1: **Adding 3-digit and 2-digit numbers (1)**

• Estimate and add 3-digit and 2-digit numbers

1 Work out the sum for each calculation.

a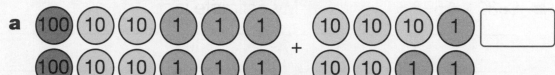

b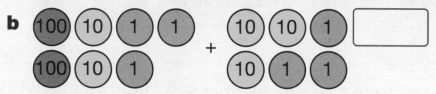

c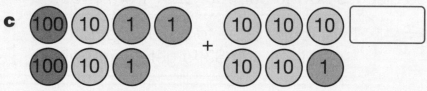

2 Look at each calculation and write an estimate.

a 316 + 72

b 483 + 19

c 624 + 41

d 736 + 53

3 Use a mental strategy to work out the answers.

a 253 + 41 =

b 342 + 35 =

c 523 + 64 =

d 712 + 54 =

e 604 + 85 =

f 444 + 33 =

Explain your strategy for **c**.

_____

_____

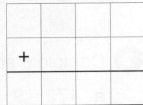

Number

4 Use the expanded written method to work out the answers.

a

	6	3	4
+		3	5

b

	7	0	4
+		7	3

c

	8	1	4
+		6	2

d

	9	0	3
+		9	1

5 Use the formal written method to work out the answers.

a 623 + 75   **b** 561 + 28   **c** 736 + 63   **d** 812 + 86

6 Jin thinks that the only way to add 704 and 35 is to use the formal written method. Do you agree? ☐ Why?

_____

_____

Show how you would work out the answer.

7 Fill in the missing numbers in these calculations.

a

	5		2
+		4	
		8	6

b

			2
+		3	
	8	7	9

c

		0	
+			6
	9	1	9

Date: _____

Number

Lesson 2: **Adding 3-digit and 2-digit numbers (2)**

• Estimate and add 3-digit and 2-digit numbers

1 Work out the sum for each calculation.

a

b

c

2 Complete each addition by adding 10 and subtracting 1.

a 172 + 9 =

b 236 + 9 =

c 687 + 9 =

d 734 + 9 =

e 564 + 9 =

f 829 + 9 =

3 Use compensation to work out the answers.

a 246 + 19 =

b 357 + 29 =

c 414 + 39 =

d 536 + 49 =

e 168 + 79 =

f 653 + 59 =

Explain compensation.

_____

_____

4 Use the expanded written method to add these numbers.

a

	6	2	8
+		4	8

b

	7	3	7
+		5	9

c

	5	4	6
+		3	7

d

	8	5	6
+		2	9

5 Use the formal written method to work out the answers.
Estimate first.

a 367 + 28   **b** 427 + 49   **c** 538 + 57   **d** 657 + 27

My estimate is [ ]

My estimate is [ ]

My estimate is [ ]

My estimate is [ ]

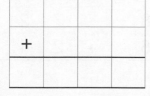

6 Three learners are making up 3-digit add 2-digit calculations.
The numbers are only used once.

629   642   658   29   48   17

a Aubrey's answer is between 640 and 650.
Which two numbers does he add together? [ ]

b Charlotte's answer is the greatest.
Which two numbers does she add together? [ ]

c Joel adds the other two numbers together.
Which numbers does he add? [ ]

What is the total? [ ]

Date: _____

Number

Lesson 3: **Adding pairs of 3-digit numbers (1)**

• Estimate and add pairs of 3-digit numbers

1 Use the making 10 strategy to find the sum of each calculation.

Decide which number you will make into a multiple of 10.

a 137 + 26 =

b 126 + 47 =

c 154 + 38 =

d 125 + 67 =

2 Use the expanded written method to work out the answers.
Estimate first.

My estimate is

My estimate is

My estimate is

My estimate is

a

	3	7	7
+	3	1	6

b

	4	3	6
+	2	4	6

c

	6	4	8
+	2	3	5

d

	7	2	5
+	2	4	6

3 Use the making 10 strategy to work out the answers. Estimate first.

a My estimate is   347 + 198 =

b My estimate is   436 + 229 =

c My estimate is   538 + 227 =

d My estimate is   636 + 148 =

Explain this strategy.

Use the formal written method to work out the answers. Estimate first.

a 427 + 384

My estimate is [ ]

b 546 + 247

My estimate is [ ]

c 426 + 457

My estimate is [ ]

d 637 + 248

My estimate is [ ]

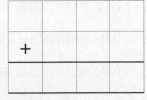

5 Joseph thinks the best way to add 449 and 238 is to use the formal written method. Patrice thinks the way to add those numbers is to use the making 10 strategy.

Who do you agree with? _____ Why?

_____

_____

Show your chosen method.

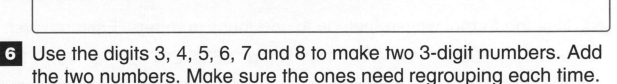

6 Use the digits 3, 4, 5, 6, 7 and 8 to make two 3-digit numbers. Add the two numbers. Make sure the ones need regrouping each time.

What other similar calculations can you make?

Date: _____

Lesson 4: **Adding pairs of 3-digit numbers (2)**

Number

• Estimate and add pairs of 3-digit numbers

You will need
• paper

1 Estimate the sum for each calculation.

a 305 + 199   My estimate is _____

b 312 + 291   My estimate is _____

c 452 + 352   My estimate is _____

d 523 + 298   My estimate is _____

e 698 + 252   My estimate is _____

2 Use the expanded written method to work out the answers.

a

	1	8	7
+	1	5	8

b

	1	8	6
+	1	4	7

c

	1	7	7
+	1	6	5

d

	2	8	5
+	1	6	8

3 Use the formal written method to add these numbers. Estimate first.

a 347 + 358
My estimate
is [ ]

b 475 + 147
My estimate
is [ ]

c 563 + 267
My estimate
is [ ]

d 487 + 267
My estimate
is [ ]

4 Sharni scores 289 points in a game. Adam scores 274. How many points do they score altogether? [ ]

5 Find the sum of 456 and 268. Make up a word problem for this calculation. Then answer the problem.
My problem

_____

_____

_____

My answer

_____

6 Samira has a collection of 387 stamps. Her grandmother gives her another 275. How many stamps does she have now? ⬚
Show how you worked this out on paper.

7 Hayden runs for 278 m. Tien runs 269 m further than Hayden.

a How many metres does Tien run? ⬚

b How many metres do they run altogether? ⬚

8 Use the digits 3, 5, 6, 7, 8 and 9 to make four addition calculations. Each of your numbers must have three different digits. The ones need regrouping each time. So do the tens.

Date: _____

☺ ☺ ☹

Number

Lesson 1: **Subtracting 2-digit from 3-digit numbers (1)**

• Estimate and subtract 2-digit numbers from 3-digit numbers

1 Use partitioning to find the difference between the numbers.

a (100)(10)(10)(1)(1)(1) – 21 [ ]
(100)(10)(10)(1)(1)(1)

b (100)(10)(10)(10)(10)(1)(1) – 42 [ ]
(100)(10)(10)(10)(1)(1)

c (100)(10)(10)(10)(10)(1)(1)(1)(1) – 67 [ ]
(100)(10)(10)(10)(10)(1)(1)(1)(1)

2 Look at each calculation and write an estimate.

a 235 – 34 [ ]   **b** 457 – 52 [ ]

c 582 – 81 [ ]   **d** 698 – 92 [ ]

3 Use a mental strategy to work out the answers.

a 456 – 34 = [ ]   **b** 569 – 48 = [ ]

c 657 – 46 = [ ]   **d** 782 – 61 = [ ]

e 874 – 52 = [ ]   **f** 395 – 73 = [ ]

Explain your strategy for **b**.

_____

_____

Number

4 Use the expanded written method to work out the answers.

Example: 658 – 36

	600	50	8
–		30	6
	600	20	2

600 + 20 + 2 = 622

a 984 – 72

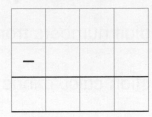

b 865 – 53

c 758 – 44

d 687 – 65

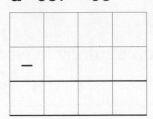

e 579 – 47

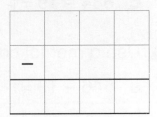

5 Use the formal written method to work out the answers.

a 456 – 35

b 576 – 32

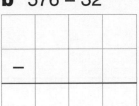

c 759 – 47

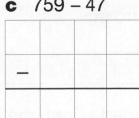

d 963 – 12

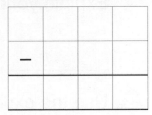

6 George, Sophie, Naomi and Tim use the digits 3, 4, 6, 7 and 8 to make a 3-digit subtract 2-digit calculation. All the digits of their calculations are different. George's calculation has a difference of 323. Sophie's calculation has a difference of 314. Naomi's calculation has a difference of 332. Tim's calculation has a difference of 341. What could their calculations be?

George [ ]   Sophie [ ]

Naomi [ ]   Tim [ ]

Date: _____

   47

Number

Lesson 2: **Subtracting 2-digit from 3-digit numbers (2)**

• Estimate and subtract 2-digit numbers from 3-digit numbers

1 Complete these subtraction calculations by subtracting 20 and adding 1.

a 432 – 19 = ☐

b 567 – 19 = ☐

c 634 – 19 = ☐

d 724 – 19 = ☐

 2 Use a mental calculation strategy to work out the answers.

a 546 – 19 = ☐

b 895 – 29 = ☐

c 657 – 39 = ☐

d 759 – 49 = ☐

e 483 – 79 = ☐

f 972 – 59 = ☐

Explain your strategy.

_____

_____

 3 Use the expanded written method to work out the answers.

Example: 896 – 38

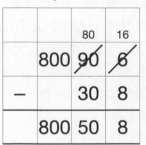

a 456 – 28

b 673 – 48

	80	16
800	~~90~~	~~6~~
–	30	8
800	50	8

800 + 50 + 8 = 858

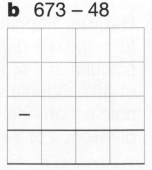

c 768 – 58

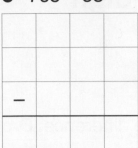

d 673 – 68

e 564 – 27

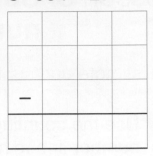

_____   _____   _____

4 Use the formal written method to work out the answers.
Estimate first.

a 657 – 39
My estimate
is [   ]

b 673 – 48
My estimate
is [   ]

c 745 – 39
My estimate
is [   ]

d 866 – 58
My estimate
is [   ]

5 Tracie, Jerome, Nancy and Tyrone use the digits 1, 3, 5, 7 and 9
to make a 3-digit subtract 2-digit calculation. All the digits of their
calculations are different.

Tracie's calculation has a difference of 958.

Jerome's calculation has a difference of 936.

Nancy's calculation has a difference of 936.

Tyrone's calculation has a difference of 918.

What could their calculations be?

Tracie [   ]   Jerome [   ]

Nancy [   ]   Tyrone [   ]

Date: _____

Number

Lesson 3: **Subtracting pairs of 3-digit numbers (1)**

• Estimate and subtract pairs of 3-digit numbers

1 Use the counting on or back strategy to find the difference. Draw a number line to help you.

a $163 - 154 =$ [ ] ⟵────────────────⟶

b $172 - 169 =$ [ ] ⟵────────────────⟶

c $186 - 169 =$ [ ] ⟵────────────────⟶

d $155 - 128 =$ [ ] ⟵────────────────⟶

e $194 - 137 =$ [ ] ⟵────────────────⟶

2 Use the expanded written method to work out the answers. Estimate first.

a $374 - 258$

My estimate is [ ]

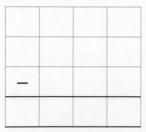

b $383 - 349$

My estimate is [ ]

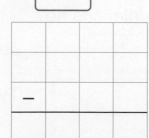

c $371 - 237$

My estimate is [ ]

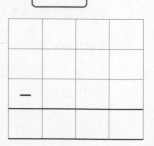

3 Use the formal written method to to work out the answers. Estimate first.

a 345 – 237

My estimate is [ ]

b 486 – 357

My estimate is [ ]

c 564 – 238

My estimate is [ ]

d 456 – 339

My estimate is [ ]

4 Joana thinks that to find the difference between 765 and 748 she can count on from 748 to 765. Ben thinks that he should use the formal written method. What would you do? Explain why.

_____

_____

_____

5 Here is one of Aja's subtraction calculations.

Is this correct? [ ] Why?

_____

_____

```
    8 6 7
 −  4 5 9
 ─────────
    4 1 2
```

6 Use the digits 2, 3, 4, 6, 8 and 9 to make four 3-digit subtract 3-digit calculations. You can only use one of each number in each calculation. The ones need regrouping each time.

Date: _____

Number

Lesson 4: **Subtracting pairs of 3-digit numbers (2)**

- Estimate and subtract pairs of 3-digit numbers

1 Estimate the difference for each calculation.

  a 559 – 521   My estimate is _____

  b 683 – 199   My estimate is _____

  c 726 – 614   My estimate is _____

  d 845 – 801   My estimate is _____

  e 915 – 503   My estimate is _____

2 Use the expanded written method to work out the answers. Estimate first.

  a 232 – 155

    My estimate is ☐

  b 238 – 16

    My estimate is ☐

  c 231 – 175

    My estimate is ☐

_____ _____ _____

3 Use the formal written method to work out the answers. Estimate first.

  a 461 – 285   **b** 534 – 277   **c** 647 – 368   **d** 714 – 467

  My estimate is ☐   My estimate is ☐   My estimate is ☐   My estimate is ☐

<table>
<tr><td>−</td><td></td><td></td></tr>
</table>

4 Nathan has $500. He spends $275.

How much money does he have left? [ ]

5 Ugne has $8.25. She spends $2.47 on a pen and $3.99 on an eraser.

How much money does she have left? [ ]

6 Use the digits 3, 4, 5, 6, 7 and 8 to make four subtraction calculations. Each of your calculations must be made from different digits. You need to make calculations that need two regroupings. Use any method you want to find the difference.

Date: _____

Lesson 1: **5 and 10 times tables**

• Understand the relationship between the 5 and 10 times tables

1 Complete these multiplication facts.

a $5 \times \boxed{} = 30$ **b** $10 \times \boxed{} = 30$ **c** $\boxed{} \times 5 = 45$

d $\boxed{} \times 10 = 70$ **e** $25 = \boxed{} \times 5$ **f** $40 = 10 \times \boxed{}$

2 Complete these division facts.

a $50 \div 5 = \boxed{}$ **b** $70 \div 10 = \boxed{}$ **c** $\boxed{} = 35 \div 5$

d $\boxed{} = 60 \div 10$ **e** $\boxed{} \div 10 = 3$ **f** $\boxed{} \div 5 = 6$

3 Ling knows that $10 \times 5 = 50$.

What commutative fact does she know? $\boxed{}$

Write the two inverse facts she also knows.

$\boxed{}$ and $\boxed{}$

4 Joseph needs to multiply 16 by 5. He decides to multiply by 10 and then double the product.

His answer is 320. Is he correct? $\boxed{}$ Explain your thinking.

_____

_____

5 Work out the answers. Use the strategy: multiply by 10 then halve.

a $12 \times 5 =$ $\boxed{}$

b $15 \times 5 =$ $\boxed{}$

Number

c 22 × 5 =

d 24 × 5 =

e 32 × 5 =

6 Mansoor knows that 130 is divisible by both 5 and 10.

How do you think he knows this?

_____

7 Hamish is thinking of a 2-digit number that is less than 100.
It is a multiple of 5 and 10.

a What number could he be thinking of?

Write all the possibilities.

b His number is in the 6 times table.

What numbers could be thinking of?

c His number has an even number of tens.

What is Hamish's number?

Date: _____

Number

Lesson 2: **2, 4 and 8 times tables**

• Understand the relationship between the 2, 4 and 8 times tables

1 Complete these multiplication facts.

a $4 \times \boxed{} = 20$ **b** $2 \times \boxed{} = 14$ **c** $\boxed{} \times 8 = 64$

d $\boxed{} \times 4 = 36$ **e** $18 = \boxed{} \times 2$ **f** $32 = 4 \times \boxed{}$

2 Complete these division facts.

a $20 \div 2 = \boxed{}$ **b** $24 \div 8 = \boxed{}$ **c** $\boxed{} = 12 \div 2$

d $\boxed{} = 16 \div 4$ **e** $\boxed{} \div 8 = 5$ **f** $\boxed{} \div 4 = 6$

3 Work out the answers. Use the strategy: multiply by 2 then double.

a $15 \times 4 =$

b $22 \times 4 =$

c $32 \times 4 =$

4 Multiply each number by 8. Use the strategy: multiply by 2 then double and double again.

a 12

b 14

c 21

d 42

Number

5 Harris knows that 6 × 8 = 48.

What commutative fact does he know?

Write the two inverse facts he also knows.

[ ] and [ ]

6 Ichiro needs to multiply 16 by 4. He decides to multiply 16 by 2 and then double the product.

His answer is 64. Is he correct? [ ] Explain your thinking.

_____

_____

7 Serena says:

> I am thinking of a 2-digit number that is less than 40. It is a multiple of 8. That means that it must also be a multiple of 2 and 4.

a Do you agree with Serena? [ ] Explain why.

_____

_____

b What numbers could she be thinking of?

c Serena says that her number is even.

Is this a helpful clue? [ ] Explain why.

_____

Date: _____

Number

Lesson 3: **3, 6 and 9 times tables**

• Understand the relationship between the 3, 6 and 9 times tables

1 Complete these multiplication facts.

a $3 \times \boxed{} = 21$ **b** $6 \times \boxed{} = 48$ **c** $\boxed{} \times 9 = 36$

d $\boxed{} \times 3 = 24$ **e** $90 = \boxed{} \times 9$ **f** $27 = 9 \times \boxed{}$

2 Complete these division facts.

a $27 \div 9 = \boxed{}$ **b** $24 \div 3 = \boxed{}$ **c** $\boxed{} = 18 \div 6$

d $\boxed{} = 45 \div 9$ **e** $\boxed{} \div 9 = 6$ **f** $\boxed{} = 18 \div 3$

3 Work out the answers. Use the strategy: multiply by 3 then double.

a $13 \times 6 =$

b $21 \times 6 =$

c $33 \times 6 =$

4 Multiply each number by 9. Use the strategy: multiply by 3 and then 3 again.

a 12

b 21

c 32

d 33

Number

5 Bernie knows that $6 \times 9 = 54$.

What commutative fact does she know?

Write the two inverse facts she also knows.

[   ] and [   ]

6 Tom needs to multiply 16 by 6. He decides to multiply 16 by 3 and then double the product.

His answer is 96. Is he correct? [   ] Explain your thinking.

_____

_____

7 Hope says:

> I am thinking of a 2-digit number. It is a multiple of 9. That means that it must also be a multiple of 3 and 6?

a Is this always, sometimes or never true? [   ]

Explain why.

_____

b Hope's number is a multiple of 3, 6 and 9.

What numbers could she be thinking of?

_____

c Hope says that her number is greater than 36 and less than 72.

What number is she thinking of? [   ] Why?

_____

Date: _____

Number

Lesson 4: **7 times table**

• Know the 7 times table

1 Complete these multiplication calculations.

a $7 \times \boxed{} = 14$ **b** $7 \times \boxed{} = 28$ **c** $\boxed{} \times 7 = 35$

d $\boxed{} \times 7 = 63$ **e** $56 = \boxed{} \times 7$ **f** $21 = 7 \times \boxed{}$

2 Complete these division calculations.

a $28 \div 7 = \boxed{}$ **b** $7 \div 7 = \boxed{}$ **c** $\boxed{} = 14 \div 7$

d $\boxed{} = 70 \div 7$ **e** $\boxed{} \div 7 = 8$ **f** $\boxed{} \div 7 = 2$

3 Use these numbers to write two multiplication and two division facts.

a 6, 7, 42

b 3, 7, 21

c 8, 7, 56

d 9, 7, 63

4 Marcus says: If I know that $7 \times 6 = 42$, I know the answer to 42 divided by 6.

Do you agree? $\boxed{}$ Explain why.

Number

 Draw a ring around the odd one out.

$4 \times 7 = 28$ $\qquad$ $7 \times 5 = 35$ $\qquad$ $2 \times 7 = 14$

$\qquad\qquad$ $6 \times 8 = 48$ $\qquad$ $7 \times 9 = 63$

Why is it the odd one out?

_____

 Matsu multiplies two numbers together. She doubles the product and ends up with 56.

Which two numbers could she have multiplied together?

Write all the possibilities.

7 **a** Adnan needs to multiply 9 by 7. He decides to multiply 9 by 4 and then 9 by 3 and add the two products together.

Explain why his strategy will work.

_____

_____

b Samira also needs to multiply 9 by 7. She decides to multiply 7 by 10 and then subtract 7.

Explain why her strategy works.

_____

_____

Date: _____

Lesson 1: **Multiples**

> • Understand and identify multiples

1 Draw a ring around the multiples of 2.

14   15   8   11   3   10   19

2 Tick the statement that is correct.

a Multiples of 2 can be a mixture of odd and even numbers. ☐

b Multiples of 2 are all even. ☐

c Multiples of 2 are all odd. ☐

3 How can you prove that 18 is a multiple of 2, 3, 6 and 9?

_____

_____

4 16 is a multiple of several numbers.

a Write the numbers 16 is a multiple of.

_____

b Write the multiplication facts to show how your numbers make 16.

_____

5 a Write six multiples of 7.

_____

b Write six multiples of 8.

_____

c Write six multiples of 9.

_____

6 Eden thinks that because 3 is an odd number, every multiple of 3 must be an odd number.

Do you agree? ☐ Explain your thinking.

_____

_____

7 Sammy thinks: ○○○◯

Do you agree? ☐

> If I add 1, 2 and 3 I get a sum of 6. That means 6 must be a multiple of 1, 2 and 3.

Why?

_____

_____

8 Mei is thinking of a 2-digit multiple of 2. Ben is thinking of a 2-digit multiple of 3. Xander is thinking of a 2-digit multiple of 4. Dom is thinking of a 2-digit multiple of 6. Alice is thinking of a 2-digit multiple of 8. They are all thinking of the same number.

a What number could they be thinking of? Write all possibilities.

```



```

b Their number is less than 30. What is it? ☐

Date: _____

Number

Lesson 2: **Factors**

• Understand and identify factors

1 Draw a ring around the factors of 15.

2   3   4   5   6   7

2 Tick the statement that shows two of the factors of 15.

a $24 \div 3 = 8$ ☐   **b** $21 \div 3 = 7$ ☐   **c** $15 \div 3 = 5$ ☐

3 How can you prove that 4 is a factor of 8, 12, 24 and 28?

_____

_____

4 4 is a factor of many numbers.

a Write five numbers that have a factor of 4.

b What do all your factors have in common?

_____

_____

5 **a** Write four factors of 12.

b Write four factors of 18.

c Write four factors of 24.

6 Jade thinks that if you multiply two factors together you will get a multiple.

Do you agree? ▢ Explain your thinking.

_____

_____

7 Logan thinks: ○○

Do you agree?

▢ Why?

> If I add 1, 2, 3 and 4 I get a sum of 10. That means 1, 2, 3 and 4 must be factors of 10.

_____

_____

8 Harrison is thinking of a number.

It has factors of 1, 2, 3, 5 and 10.

His number is the smallest one possible with these factors.

a What number is he thinking of? ▢

b What other factors does his number have?

▢

9 Naomi says:

> I think that the sum of two even 1-digit numbers is going to be a multiple of 4.

Is this always, sometimes or never true? ▢
Explain why.

_____

_____

_____

Date: _____

Number

Lesson 3: **Multiples and factors**

• Understand the relationship between multiples and factors

1 Draw a ring around the multiples of 7.

14   30   42   63   81

2 Draw a ring around the multiples of 8.

12   80   32   28   64

 3 Fill in the boxes to make these calculations correct.

a ☐ × 6 = 42   **b** ☐ × 5 = 40   **c** ☐ × 8 = 24

42 ÷ ☐ = 7   40 ÷ ☐ = 8   24 ÷ ☐ = 3

d ☐ × 6 = 60   **e** ☐ × 9 = 27   **f** ☐ × 7 = 14

60 ÷ ☐ = 10   27 ÷ ☐ = 3   14 ÷ ☐ = 2

4 a Draw a ring around the factors of 50.

25   10   7   9   5

b Draw a ring around the factors of 40.

8   5   3   9   12   1

c Draw a ring around the factors of 35.

8   9   7   5   10

5 Draw a ring around the multiples of both 4 and 5.

20   25   30   35   40

6 Put the words factor and multiple in the boxes so the statements are correct.

☐ × ☐ = ☐

☐ ÷ ☐ = ☐

Choose your own numbers to write in the boxes to make the statements correct.

☐ × ☐ = ☐   ☐ ÷ ☐ = ☐

7 Tai needs to multiply 16 by 6. He decides to multiply 16 by 3 and then double the product.

His answer is 96. Is he correct? ☐ Explain your thinking.

_____

_____

8 Hope says: I am thinking of a 2-digit number. It is a multiple of 9. That means that it must also be a multiple of 3 and 6?

a Is this always, sometimes or never true? ☐ Explain why.

_____

_____

b Hope's number is a multiple of 3, 6 and 9.

What numbers could she be thinking of?

_____

c Hope says that her number is greater than 36 and less than 72.

What number is she thinking of? ☐ Why?

_____

Date: _____

Number

Lesson 4: **Tests of divisibility**

• Understand tests of divisibility by 2, 5, 10, 25, 50 and 100

1 Draw a ring around the numbers that are divisible by 2.

18   65   70   179   221   334

2 Draw a ring around the numbers that are divisible by 5.

45   60   98   145   205   436

3 Draw a ring around the numbers that are divisible by 10.

67   90   165   280   341   470

4 Draw a ring around the odd one out.

30   80   125   250   300

Why is it the odd one out?

_____

5 Look at this table.

25	50	75	100	125	150	175	200	225	250
50	100	150	200	250	300	350	400	450	500
100	200	300	400	500	600	700	800	900	1000

Write eight numbers that are divisible by 2, 5 and 10.

| |
| |
| |
| |

6 Brad says:

> All numbers that can be divided by 10 can also be divided by 2 and 5.

Do you agree? [   ] Explain why.

_____

Number

7   Suzi says: I think 125, 250 and 375 are divisible by 5.

Do you agree? ☐   Explain why.

_____

_____

8   Write eight numbers that have a factor of 25.

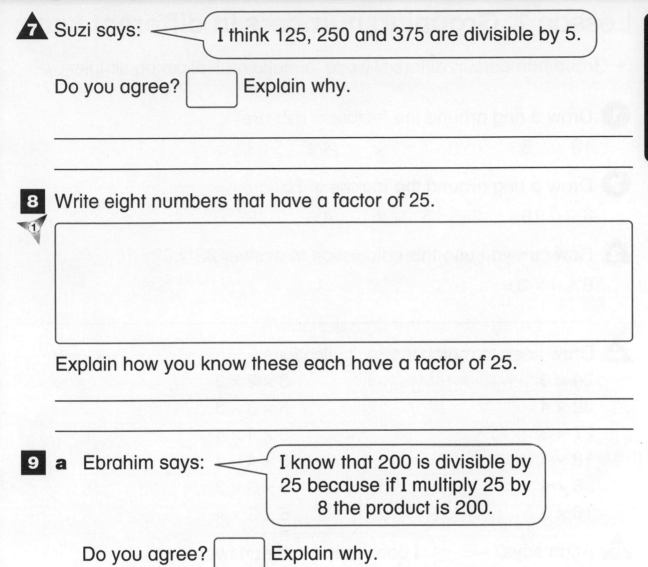

Explain how you know these each have a factor of 25.

_____

_____

9   **a**   Ebrahim says: I know that 200 is divisible by 25 because if I multiply 25 by 8 the product is 200.

Do you agree? ☐   Explain why.

_____

_____

b   Sasha says: I know that 200 is divisible by 25 because if I divide 200 by 25 the quotient is 8.

Do you agree? ☐   Explain why.

_____

_____

Date: _____

Number

Lesson 1: **Grouping numbers in different ways**

• Group numbers in different ways to make multiplication simpler

1 Draw a ring around the factors of 12.

10   6   8   7   3   4   5

2 Draw a ring around the factors of 20.

6   10   2   3   5   4   7

3 How can you use this calculation to answer 32×3?

$8 \times 4 \times 3 =$

_____

4 Draw lines to match the multiplications.

24×3	$8 \times 2 \times 3$
32×4	$8 \times 3 \times 3$
27×3	$9 \times 4 \times 4$
18×4	$8 \times 4 \times 4$
36×4	$9 \times 3 \times 3$
16×3	$6 \times 3 \times 4$

5 Arlan says:

> I can work out the answer to 18×6 by using factors of 18.

This is what he does: $9 \times 2 \times 6 = 54 \times 2 = 108$

Explain what he has done.

_____

_____

6 Draw a ring around the mistake in each calculation. Write the correct calculation.

a $36 \times 3 = 9 \times 5 \times 3$

_____

b $28 \times 4 = 9 \times 4 \times 4$

_____

Number

c $45 \times 4 = 9 \times 6 \times 4$

_____

7   **a**   Sami finds the factors of a 2-digit number that he has to multiply by 6.

     These are his factors: 2, 3, 4.

     What was his original calculation? ☐

     Use Sami's method to work out the product.

     ┌───┐
     │   │
     └───┘

     Explain what he did.

_____

_____

  b   Tyra finds the factors of a 2-digit number that she has to multiply by 8.

     These are her factors: 2, 4, 6.

     What was her original calculation? ☐

     Use Tyra's method to work out the product.

     ┌───┐
     │   │
     └───┘

     Explain what she did.

_____

_____

Date: _____

Number

Lesson 2: **Multiplying tens**

- Multiply tens numbers by 1-digit numbers

1 Complete these 10 times table facts.

a 10 × 1 = ☐   **b** 10 × 2 = ☐   **c** 10 × 3 = ☐

d 10 × 4 = ☐   **e** 10 × 5 = ☐   **f** 10 × 6 = ☐

g 10 × 7 = ☐   **h** 10 × 8 = ☐   **i** 10 × 9 = ☐

2 How can you use **1** to help you find the first 10 facts in the 20 times table?

_____

3 **a** How can you multiply 80 by 7?

☐

b How can you multiply 50 by 8?

☐

c How can you multiply 60 by 5?

☐

4 What multiplication does this show?

☐

Write the commutative fact and the two inverse division facts.

5 Fill in the missing numbers on this number line.

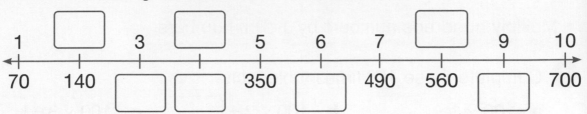

```
   [  ]      3    [  ]    5    6    7    [  ]    9    10
1
70    140    [  ]  [  ]   350  [  ]  490  560  [  ]   700
```

6 Danni and Bertie are multiplying 10s numbers by 1-digit numbers.

a Danni says: ◁— The answer to my question is 360. What is my question?

Write three possible questions.

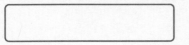   [ ]   [ ]

b Bertie says: ◁— The answer to my question is 240. What is my question?

Write three possible questions.

7 Choose pairs of numbers to multiply together and then multiply by 10.

How many different products can you make?

| 7 | 9 | 4 | 6 |

Show the calculations for your different products.

Date: _____

Number

Lesson 3: **Multiplying hundreds**

• Multiply hundreds numbers by 1-digit numbers

1 Complete these 100 times table facts.

a $100 \times 1 = \boxed{}$ **b** $100 \times 2 = \boxed{}$ **c** $100 \times 3 = \boxed{}$

d $100 \times 4 = \boxed{}$ **e** $100 \times 5 = \boxed{}$ **f** $100 \times 6 = \boxed{}$

g $100 \times 7 = \boxed{}$ **h** $100 \times 8 = \boxed{}$ **i** $100 \times 9 = \boxed{}$

2 How can you use **1** to help you find the first 10 facts in the 200 times table?

_____

3 **a** How can you multiply 800 by 7?

$\boxed{}$

b How can you multiply 500 by 8?

$\boxed{}$

c How can you multiply 600 by 5?

$\boxed{}$

4 What multiplication does this show?

$\boxed{}$

Write the commutative fact and the two inverse division facts.

Number

5 Fill in the missing numbers on this number line.

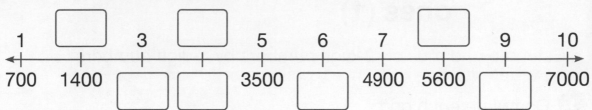

| 1 | | 3 | | 5 | 6 | 7 | | 9 | 10 |

700   1400   3500   4900   5600   7000

6 Serge and Sara are multiplying 100s numbers by 1-digit numbers.

a Serge says: The answer to my question is 2000. What is my question?

Write three possible questions.

b Sara says: The answer to my question is 3600. What is my question?

Write three possible questions.

7 Sophie thinks that when multiplying by 100, all you need to do is add two zeros.

Write an explanation for Sophie so that she can understand what happens when multiplying by 100.

Show four examples.

Date: _____

75

Lesson 4: **Multiplying 2-digit numbers and ones (1)**

- Estimate and multiply 2-digit numbers by 1-digit numbers

1 Complete each grid.

a

×	40	2
3		

42 × 3 = ☐

b

×	50	6
4		

56 × 4 = ☐

2 Use partitioning to answer each calculation.

a 52 × 6 = ☐

b 78 × 4 = ☐

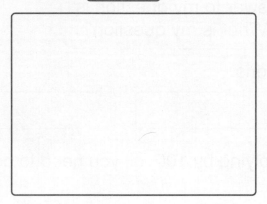
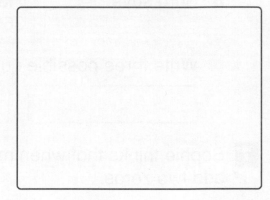

3 Using only the digits 4, 7 and 8, write six different 2-digit × 1-digit calculations. Then use any method to work out the answers.

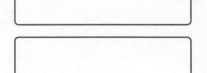
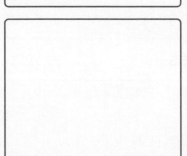

4 Tick the calculations that are correct. Cross those that are incorrect.

a $42 \times 4 = 168$ [ ]   **b** $73 \times 5 = 355$ [ ]

c $69 \times 3 = 207$ [ ]   **d** $47 \times 5 = 235$ [ ]

e $94 \times 6 = 544$ [ ]   **f** $48 \times 7 = 336$ [ ]

5   **a**   Make the target number of 195 using three of these digits.

2   3   4   5   6   7   [ ] [ ] $\times$ [ ]

b   Use the remaining three digits to make another calculation with a product less than 150.

c   Use them again to make another calculation with a product greater than 150.

Date: _____

Lesson 1: **Multiplying 2-digit numbers and ones (2)**

- Estimate and multiply 2-digit numbers by 1-digit numbers

1 a What multiplication calculation does this show?

(10) (10) (10) (1) (1)
(10) (10) (10) (1) (1)
(10) (10) (10) (1) (1)

b What is the product?

2 Complete the grid.

$34 \times 3 =$

×	30	4
3		

3 Use the expanded written method to answer each calculation.

a

		4	9
×			3
+			

☐ × ☐
☐ × ☐

b

		5	8
×			5
+			

☐ × ☐
☐ × ☐

c

		6	7
×			3
+			

☐ × ☐
☐ × ☐

d

		7	3
×			4
+			

☐ × ☐
☐ × ☐

4 Using only the digits 5, 6 and 7, write four different 2-digit × 1-digit calculations. Then use the expanded written method to work out the answers.

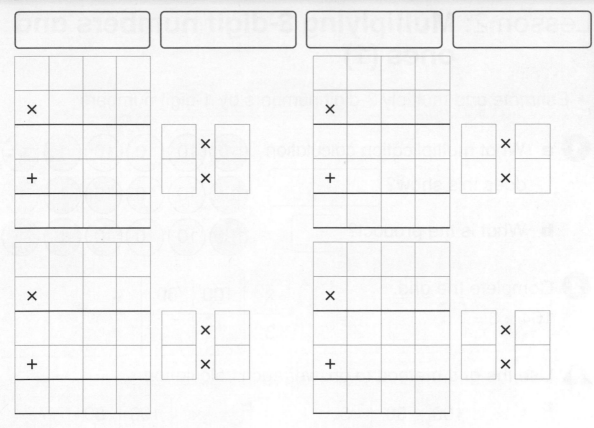

5 Rhiannon has been shown how to work out the product of two numbers using the expanded written method. She does it in two different ways.

		6	3
×			7
	4	2	0
+		2	1
	4	4	1

		6	3
×			7
		2	1
+	4	2	0
	4	4	1

a What is the same about both calculations?

_____

b What is different?

_____

c Is one method better than the other? Explain why.

_____

_____

Date: _____

Number

Lesson 2: **Multiplying 3-digit numbers and ones (1)**

• Estimate and multiply 3-digit numbers by 1-digit numbers

1 **a** What multiplication calculation does this show? ☐

      100 10 10 10 1 1
      100 10 10 10 1 1
      100 10 10 10 1 1

b What is the product? ☐

2 Complete the grid.

$164 \times 3 =$ ☐

×	100	60	4
3			

3 Use the grid method to answer each calculation.

a

×	100	30	5
4			

$135 \times 4 =$ ☐

b

×	100	80	7
3			

$187 \times 3 =$ ☐

4 Use the grid method to answer each calculation.

a $154 \times 4 =$ ☐

b $163 \times 5 =$ ☐

c $128 \times 6 =$ ☐

d $136 \times 7 =$ ☐

5 Aki and Ekon have completed the same multiplication.

Aki   Ekon

×	100	60	4
5	500	300	20

×	100	60	4
5	500	30	20

$500 + 300 + 20 = 820$ $500 + 30 + 20 = 550$

a   Who has the correct product?

b   What is wrong with the other answer?

_____

_____

6 Georgie is using the grid method to complete this multiplication.

×			
	800	80	8

She thinks there are three possible calculations.

She is correct.

What are the three possible calculations?

Show them in the grids.

Solution 1   Solution 2

×			
	800	80	8

×			
	800	80	8

Solution 3   Working out

×			
	800	80	8

Date: _____

Number

Lesson 3: **Multiplying 3-digit numbers and ones (2)**

• Estimate and multiply 3-digit numbers by 1-digit numbers

You will need
• paper

1 **a** What multiplication calculation does this show?

b What is the product?

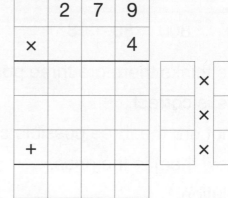

2 Estimate the answer to each calculation, writing your estimate in the bubble. Then use the expanded written method to work out the answer.

a

	2	6	8
×			3
+			

×
×
×

b

	2	7	9
×			4
+			

×
×
×

c

	2	3	4
×			3
+			

×
×
×

d

	2	5	6
×			4
+			

×
×
×

3 Nisha and Malachi both multiply 148 by 5.

Nisha says: I multiply 148 by 10 and halve it to get 740.

Malachi says: I multiply 100 by 5, 40 by 5 and 8 by 5. Then I add 500, 200 and 40 to get 740.

Whose method would you use? [   ] Why?

_____

_____

4 Naomi uses partitioning to calculate 245×3.

This is what she does:

Her answer is 6135.

$$245 \times 3 =$$
$$200 \times 3 \qquad 40 \times 3 \qquad 5 \times 3$$
$$600 \quad + \quad 120 \quad + \quad 15$$

a Explain the mistake she has made.

_____

_____

b What is the correct product? [   ]

5 Use the digit cards to make 3-digit numbers. ｜ 3 ｜ 1 ｜ 2 ｜

Multiply your numbers by 3.

Use paper to record your working out.

a What is the greatest product you can make? [   ]

What was your calculation? [   ]

b What is the smallest product you can make? [   ]

What was your calculation? [   ]

Date: _____

Number

Lesson 4: **Multiplying 3-digit numbers and ones (3)**

• Estimate and multiply 3-digit numbers by 1-digit numbers

1 Complete the multiplication calculations.

a

	3	2	7
×			3
+			

	×	
	×	
	×	

b

	2	2	4
×			4
+			

	×	
	×	
	×	

c

	1	3	3
×			5
+			

	×	
	×	
	×	

d

	1	3	4
×			6
+			

	×	
	×	
	×	

2 Complete the grids.

a

×	300	20	4
3			

$324 \times 3 = \boxed{}$

b

×	100	60	5
3			

$165 \times 3 = \boxed{}$

3 Sophie has 234 coins. Tom has 3 times as many. How many coins does Tom have? Use partitioning to find out.

4 Daisy and Alton are calculating 234 × 2. Daisy decides to double. This gives her a product of 468. Alton decides to use the grid method. He also gets a product of 468.

Which method would you use? _____

Why?

_____

_____

5 Masa walks 245 m home from school. Billy walks 3 times as far.

How far does Billy walk? ☐ Show your working.

6 Sian and Lisa are taking part in a reading competition. One morning Sian reads 164 pages of her book. Lisa reads 4 times as many pages.

a How many pages do they read altogether? ☐

b How many fewer pages does Sian read than Lisa? ☐

Date: _____ ☺ ☐ ☹

Number

Lesson 1: **Dividing 2-digit numbers by 2 and 4**

• Estimate and divide 2-digit numbers by 2 and 4

1 Use partitioning and halving to divide each number by 2. Show your working.

a 36

b 48

c 64

d 58

2 Use partitioning and halving twice to divide each number by 4. Show your working.

a 48

b 52

c 64

d 56

3 Think of four numbers that will have a remainder of 3 when divided by 4. Divide your numbers by 4 to see if you are correct.

a

b

c

d

Number

4 Explain why there will be a remainder of 1 when you divide an odd number by 2. _____

_____

5 Use partitioning and halving twice to divide each number by 4. Make sure you write the remainders. Show your working.

a 58

b 71

c 83

d 97

6 Sophie says:

All odd numbers will have a remainder if divided by 2 and 4.

Is this always, sometimes or never true? Explain why.

_____

_____

7 **a** What type of number can give a remainder of 1 when divided by 4?

_____

_____

b What type of number can give a remainder of 2 when divided by 4?

_____

_____

c What type of number can give a remainder of 3 when divided by 4?

_____

_____

Date: _____

Lesson 2: **Dividing 2-digit numbers by 5**

Number

• Estimate and divide 2-digit numbers by 5

1 **a** $45 \div 5 = \boxed{}$ **b** $20 \div 5 = \boxed{}$ **c** $15 \div 5 = \boxed{}$

d $30 \div 5 = \boxed{}$ **e** $25 \div 5 = \boxed{}$ **f** $35 \div 5 = \boxed{}$

2 Divide each number by 5 by dividing by 10 and doubling.

a $60 \div 5 =$

b $75 \div 5 =$

c $85 \div 5 =$

d $90 \div 5 =$

3 Use partitioning and tables facts to divide each number by 5.

a $55 \div 5 =$ (55)

b $95 \div 5 =$ (95)

 $+$ $\boxed{}$ $=$ $\boxed{}$

 $+$ $\boxed{}$ $=$ $\boxed{}$

4 Find the quotients. Use partitioning. Remember the remainders!

a $64 \div 5 =$

b $61 \div 5 =$

c $77 \div 5 =$

d $84 \div 5 =$

5 Zane says:

Do you agree?

$\boxed{}$

I think that when I divide a number that has 4 in the ones position by 5, there will always be a remainder of 4.

Number

Why? _____

_____

6 Sam and Trisha both answered: 67 ÷ 5 =.

Sam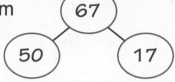

Trisha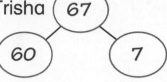

Sam: 67, 50, 17

| 10 | + | 3 r 2 | = | 13 r 2 |

Trisha: 67, 60, 7

| 10 | + | 1 r 2 | = | 11 r 2 |

Who has the correct quotient? [   ]

What is wrong with the other answer?

_____

_____

7 Sabina, Tim and Chloe are talking about dividing by 5.

Sabina says: There will always be a remainder of 3 if a number has an 8 in the ones position.

Tim says: There will also be a remainder of 3 if a number has a 3 in the ones position.

Chloe says: I think there will be a remainder of 6 if a number has a 6 in the ones position.

a Write the names of the learners who are correct.

_____

b Who is not correct? _____

c Why is that learner incorrect?

_____

_____

Date: _____

Lesson 3: **Dividing 2-digit numbers by 3 and 6**

Number

• Estimate and divide 2-digit numbers by 3 and 6

1 Divide each number by 3.

a 27 ☐   b 12 ☐   c 18 ☐

d 24 ☐   e 9 ☐   f 21 ☐

2 Divide each number by 6.

a 24 ☐   b 18 ☐   c 36 ☐

d 54 ☐   e 30 ☐   f 42 ☐

3 Use partitioning and times tables facts to divide each number by 3.

a （69）

☐ + ☐ = ☐

b （78）

☐ + ☐ = ☐

c （48）

☐ + ☐ = ☐

d （57）

☐ + ☐ = ☐

4 Find the quotient. Use partitioning.

a 78 ÷ 6 =

b 72 ÷ 6 =

c 84 ÷ 6 =

d 90 ÷ 6 =

Number

5 Molly and Zachariah both answered: 72 ÷ 6 =.

Molly says: I use partitioning and times tables facts.

Zachariah says: I divide by 3 and halve my answer.

a Which method would you use?

b Why?

_____

_____

c Use your preferred method to calculate 96 ÷ 6 =.

6 Zara says: I am dividing by 3. I think that all the numbers that are one more than a multiple of 3 will have a remainder of 1.

a Do you agree?

b Write three examples of numbers that will give a remainder of 1 when you divide them by 3.

7 **a** Use the digits 6, 7 and 8 to make 2-digit numbers.

Write the 2-digit numbers in the box.

b Divide each of your numbers by 3.

Use partitioning and times tables facts.

Date: _____

91

Number

Lesson 4: **Dividing 2-digit numbers by 7, 8 and 9**

• Estimate and divide 2-digit numbers by 7, 8 and 9

1 Divide each number by 7. Don't forget the remainders.

a 65 ☐   **b** 73 ☐   **c** 26 ☐

d 17 ☐   **e** 58 ☐   **f** 50 ☐

2 Divide each number by 8. Don't forget the remainders.

a 65 ☐   **b** 73 ☐   **c** 26 ☐

d 17 ☐   **e** 58 ☐   **f** 85 ☐

3 Divide each number by 9. Don't forget the remainders.

a 76 ☐   **b** 85 ☐   **c** 94 ☐

d 67 ☐   **e** 49 ☐   **f** 50 ☐

4 Use partitioning and times tables facts to divide each number by 7.

a

86

☐ + ☐ = ☐

b

93

☐ + ☐ = ☐

c

88

☐ + ☐ = ☐

d

97

☐ + ☐ = ☐

5 Frankie and Maisie have to calculate $81 \div 7 =$.

Frankie decides to partition. He makes 70 and 11 and uses times tables facts. This gives him a quotient of 11 r 4.

Maisie just uses tables facts. She knows that $11 \times 7 = 77$. This gives her the same quotient of 11 r 4.

Which method would you use? _____

Why?

_____

_____

6 Mi-Cha calculates $97 \div 8 =$.

She partitions 97 into 90 and 7.

She knows that $90 \div 8 = 11$ r 2 and $7 \div 8$ cannot be done so there is another 7 left over.

She makes the quotient 11 r 9.

She checks her answer by multiplying 8 by 11 and adding 9. This gives her 97.

She thinks she must be right.

a What mistake has she made?

_____

_____

b What should the quotient be? ☐

7 Abrahim thinks that because he knows all the times tables facts to 10×10, he can use these to divide any 2-digit number by 9.

Do you agree? ☐
Explain why.

_____

_____

Date: _____

Number

Lesson 1: **Working towards a written method**

• Estimate and divide 2-digit numbers by 1-digit numbers

1 **a** Draw loops around groups of 3.

(10) (10) (10) (1) (1) (1) (1) (1) (1)

What calculation have you carried out? ☐

b Draw loops around groups of 4.

(10) (10) (10) (10) (1) (1) (1) (1) (1) (1) (1) (1)

What calculation have you carried out? ☐

2 Use place value counters to work out each calculation.
Draw a diagram and explain what you did.

a 42 ÷ 3 = ☐

b 48 ÷ 3 = ☐

3 David uses place value counters to answer 63 ÷ 5 =.

This is what he says:

> I made one group of 5 tens.
> I have 4 counters left.
> That means 63 ÷ 5 is 1 remainder 4.

a Is he correct?

b Show David what he should have done with the place value counters.

c Now write an explanation so that he understands how to divide.

_____

_____

4 Abel and Jodie are checking Jack's division method.

This is what Jack does:

84 ÷ 3 =

Abel says:

> This is correct because there are columns of 3 counters.

Jodie says:

> This is not correct because Jack has made groups of 4. He should have made groups of 3.

Who do you agree with?   Why?

_____

_____

Date: _____

95

Lesson 2: **Written method of division (1)**

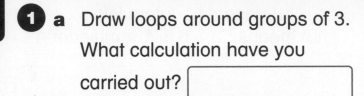

- Estimate and divide 2-digit numbers by 1-digit numbers using a written method

You will need
- paper

1 **a** Draw loops around groups of 3. What calculation have you carried out?

10s	1s
X X X	X X X
X X X	

b Draw loops around groups of 4. What calculation have you carried out?

10s	1s
X X X X	X X X X
	X X X X

2 **a** Draw a diagram to show how you can divide 57 by 4.

10s	1s

Write the calculation using the expanded written method.

b Do the same for 76 ÷ 3 =.

10s	1s

Number

3 Hayden draws this to answer 79 ÷ 3 =.

This is what he says:

> I made two groups of 3 tens.
> Then I made three groups of 3 ones.
> That means 79 ÷ 3 is 23.

10s	1s
Ⓧ X X	Ⓧ X X
Ⓧ X X	Ⓧ X X
X	Ⓧ X X

Is he correct? ☐

Show Hayden what he should have done in this place value grid.

10s	1s

4 Samira is working out 96 ÷ 4 =.

Before she starts, she says:

> I don't need to use a written method or draw. I know that the quotient will be 24.

Do you agree? ☐

What do you think she did?

_____

_____

5 Cai says:

> These dividends are all multiples of 3. So, there will be no remainders if I divide them by 3.

69 ÷ ☐ = ☐   72 ÷ ☐ = ☐   54 ÷ ☐ = ☐

Do you agree? ☐ Divide them by 3 to check if you are correct.

Use paper to show your working.

6 Use paper to work out the answer to each calculation. First **estimate** the answer, then use the **expanded written method**.

a 97 ÷ 8 = ☐   **b** 98 ÷ 7 = ☐   **c** 71 ÷ 4 = ☐

Date: _____

Number

Lesson 3: **Written method of division (2)**

• Estimate and divide 2-digit numbers by
1-digit numbers using a written method

1 Find the quotient. Don't forget the remainder.

a $45 \div 6 =$ ⬜

b $20 \div 7 =$ ⬜

c $15 \div 8 =$ ⬜

d $30 \div 9 =$ ⬜

e $25 \div 6 =$ ⬜

f $31 \div 4 =$ ⬜

2 Complete these calculations.

a ⬜ r ⬜
```
  6 ) 7 6
  –   6 0
        1 6
  –     1 2
      ┌─────┐
      └─────┘
```

b ⬜ r ⬜
```
  6 ) 7 9
  –   6 0
        1 9
  –     1 8
      ┌─────┐
      └─────┘
```

c ⬜ r ⬜
```
  7 ) 8 1
  –   7 0
        1 1
  –       7
      ┌─────┐
      └─────┘
```

3 a Explain how you divide 92 by 8.

_____

_____

_____

Write the calculation using the
expanded written method.

b Explain how you divide 89 by 7.

_____

_____

_____

Write the calculation using the
expanded written method.

Number

4 Find the quotients. Use the expanded written method.

a 73 ÷ 6 =

b 79 ÷ 6 =

c 83 ÷ 7 =

d 85 ÷ 7 =

5 Benji works out that 85 ÷ 8 = 10 r 5.

He says: 85 must be 5 away from a multiple of 8.

Do you agree? ☐ Why?

_____

_____

6 Bobby and Sunita have completed the same division.

Tick the correct answer.

What is wrong with the other answer?

Bobby

```
    1 2 r 3
  7 ) 8 7
  -  7 0
     1 7
  -  1 4
        3
```

Sunita

```
      1 1
  7 ) 8 7
  -  8 0
        7
  -     7
        0
```

_____

_____

_____

_____

Date: _____

99

Number

Lesson 4: **Using division to solve problems**

- Estimate and divide 2-digit numbers by 1-digit numbers to solve problems

1 a A farmer collects 24 eggs. He puts them in boxes of 6.

How many boxes does he fill?

b How many boxes will he fill if he collects 36 eggs?

c How many boxes will he fill if he collects 42 eggs?

d How many boxes will he fill if he collects 54 eggs?

2 a 24 learners are going on a school trip. They are travelling in minibuses. 8 learners are allowed on one minibus.

How many minibuses are needed?

b How many are needed if there are 32 learners?

c How many are needed if there are 48 learners?

d How many are needed if there are 64 learners?

3 Hiro and Kam pick 49 flowers. They put 3 flowers in each vase.

How many vases do they use?

How many flowers are left?

How did you work that out?

4 **a** Jayden buys 52 toy wheels. He makes toy cars with 4 wheels. How many toy cars can he make? ☐

Show how you worked this out.

b How many can he make if he buys 64 toy wheels? ☐

Show how you worked this out.

c How many can he make if he buys 96 toy wheels? ☐

Show how you worked this out.

5 Tia makes 96 cakes. She thinks that she will need more plates if she puts 8 on a plate than if she puts 6 on a plate. She thinks this because 8 is more than 6.

Do you agree? ☐ Why?

_____

_____

_____

6 Make up a problem for 56 ÷ 4. Then solve it.

Date: _____

Lesson 1: **Understanding place value (1)**

Number

• Understand the value of each digit in a 4-digit number

1 Write the number shown on each abacus.

a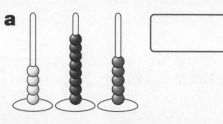

b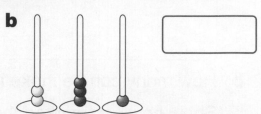

c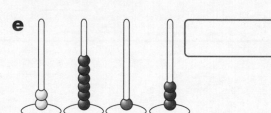

d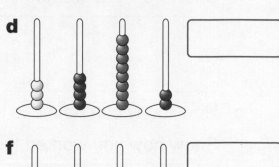

e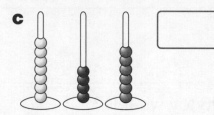

f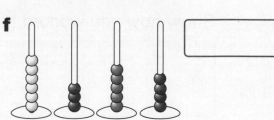

2 Represent these 4-digit numbers in any way you like.

a 2143

b 2243

c 2343

d What is the same about these numbers?

_____

e What is different?

_____

Number

3 Write each number represented by the place value counters. Then write the letter of the description that matches.

| **A** The greatest number | **B** A number where the 3 digit is worth three 1000s |

| **C** A multiple of 100 | **D** A number with an odd number of 10s |

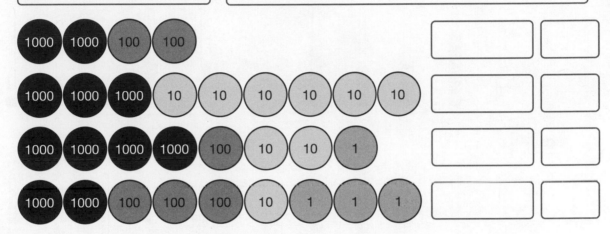

4 What is the value of each digit?

a 2634 = ☐ + ☐ + ☐ + ☐

b 5217 = ☐ + ☐ + ☐ + ☐

c 8262 = ☐ + ☐ + ☐ + ☐

5 You have eight beads.

Use all the beads to make these numbers.

a The greatest 4-digit number.

b The smallest 4-digit number.

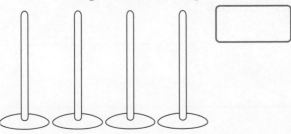

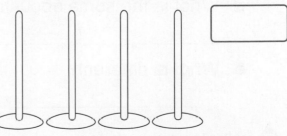

Date: _____

Number

Lesson 2: **Understanding place value (2)**

• Understand the value of each digit in a 5-digit number

1 Write the number shown on each abacus.

a [ ]   **b** [ ]

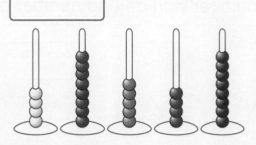   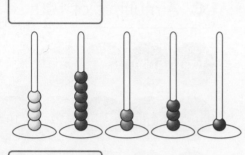

c [ ]   **d** [ ]

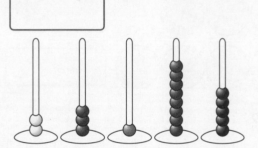   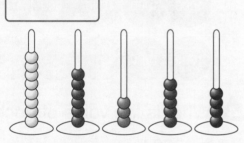

2 Represent these 5-digit numbers in any way you like.

a 21 943   **b** 22 943   **c** 23 943

d What is the same about these numbers?

_____

e What is different?

_____

3 What is the value of each digit?

a 23 456 = [ ] + [ ] + [ ] + [ ] + [ ]

104

Number

b 41 893 = ☐ + ☐ + ☐ + ☐ + ☐

c 67 275 = ☐ + ☐ + ☐ + ☐ + ☐

d 39 126 = ☐ + ☐ + ☐ + ☐ + ☐

4 Mahmood has made 24 000. He uses the digits 5, 7 and 8 and puts them in the 100s, 10s and 1s positions. He makes 24 578, 24 785 and 24 857. He thinks that he has made all the

different possibilities. Do you agree? ☐ Explain why.

_____

_____

5 Heena says:

> I have used place value counters to represent 12 345.

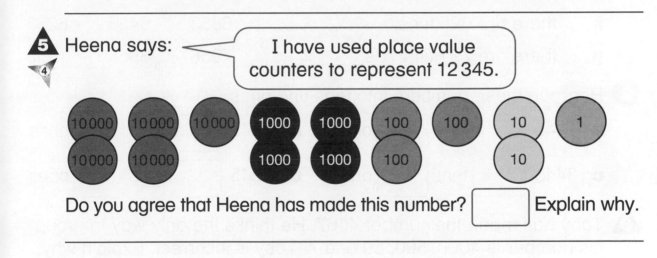

Do you agree that Heena has made this number? ☐ Explain why.

_____

_____

6 You have nine beads.

Using all nine beads each time, make three different 5-digit numbers. Each number must have 4 hundreds and 3 tens.

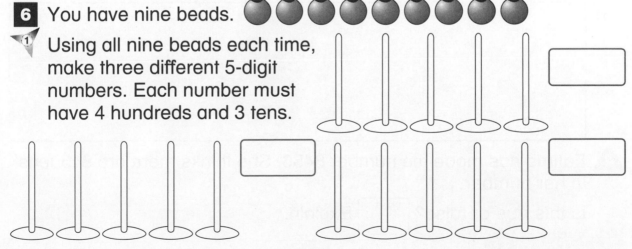

Date: _____

Number

Lesson 3: **Regrouping**

• Regroup 4-digit numbers in different ways

1 Draw a ring around the number in which…

a	…there are 8 ones	1345	1348	1349
b	…there are 9 tens	4394	4374	4324
c	…there are 3 hundreds	5682	5482	5382
d	…there are 4 thousands	1546	4546	7546
e	…there are 1327 ones	1324	1325	1327
f	…there are 625 tens	6853	6453	6253
g	…there are 79 hundreds	7956	3956	1956

2 Regroup these numbers into tens and ones.

a 1245 ☐ tens ☐ ones   **b** 2345 ☐ tens ☐ ones

c 3445 ☐ tens ☐ ones   **d** 4545 ☐ tens ☐ ones

3 Toby has made the number 4567. He thinks the only way to group his number is 4000, 500, 60 and 7. Toby is incorrect. Explain why, with six examples to prove your thinking.

4 Fatima has made the number 8456. She thinks there are 845 tens in her number.

Is this true or false? ☐ Explain.

5 Majid regrouped a number.

He made 1 thousand, 6 hundreds, 36 tens and 15 ones.

What was his number? ⬚

6 Write a 4-digit number that fits the clues.

a An odd 4-digit number with 47 ones. ⬚

b A number with 35 hundreds. ⬚

c An even number with 57 tens. ⬚

7 Megan and Olivia both have these digit cards.

| 5 | 3 | 0 | 7 | 8 |

a What is the greatest 4-digit number that Megan can make with these digits where there are 53 hundreds? ⬚

b What is the smallest 4-digit number that Megan can make with these digits where there are 35 tens? ⬚

c What number can she make where there are 805 tens? ⬚

d Olivia uses four of the digit cards to make 8073. Write one way she can regroup her number.

⬚

Can you think of any other ways? Write three more.

_____

_____

_____

Date: _____ 🙂 😐 ☹

Number

Lesson 4: **Comparing and ordering numbers**

• Compare and order numbers, including negative numbers

1 Write > or < in between each pair of numbers.

a 459 ☐ 759   **b** 302 ☐ 802   **c** 644 ☐ 244

d 592 ☐ 992   **e** 420 ☐ 120   **f** 737 ☐ 837

2 Order the numbers by marking them on each number line.

a 550   350   750

← 0 100 200 300 400 500 600 700 800 900 1000 →

b 620   290   180

← 0 100 200 300 400 500 600 700 800 900 1000 →

c −1   7   −9

← −10 −8 −6 −4 −2 0 2 4 6 8 10 →

3 Write > or < between each pair of numbers.

a 2580 ☐ 5208   **b** 1252 ☐ 5212   **c** 6933 ☐ 3936

d 4867 ☐ 8467   **e** 1569 ☐ 1596   **f** 7110 ☐ 7101

4 For each pair of numbers in **3**, write another 4-digit number that could go between them.

a ☐   **b** ☐   **c** ☐

d ☐   **e** ☐   **f** ☐

Number

5 Mark where you estimate the numbers lie on each number line.

a 25   −22   8   −37   −7

−40   40

b 385   358   362   886   680   608

−40 line

300   1000

c 4219   6380   6308   4921   2977

2000   10 000

d 3053   8071   3503   2789   2987

2000   10 000

6 Petra says: I have ordered some numbers.

This is her order: 23   −45   57   −102   148   −200

Do you agree with her? ☐ Explain why.

_____

_____

7 Complete each list so that the 4-digit numbers are in order,
smallest first. You can only use each digit, 0–9, once in a list.

| 0 | 1 | 2 | 3 | 4 | 5 | 6 | 7 | 8 | 9 |

a 1 8 ☐ ☐ , 1 8 2 ☐ , 1 8 ☐ ☐ , 1 8 ☐ ☐ , 1 8 ☐ ☐

b 7 5 ☐ ☐ , 7 5 ☐ ☐ , 7 5 ☐ ☐ , 7 5 7 ☐ , 7 5 ☐ ☐

c ☐ ☐ 4 5, ☐ ☐ 9 0, 5 1 ☐ ☐ , 5 ☐ ☐ ☐

Date: _____

Number

Lesson 1: **Multiplying and dividing by 10**

- Use place value to multiply and divide whole numbers by 10

1 Multiply each number by 10.

  a 42 ☐   **b** 13 ☐   **c** 56 ☐

  d 73 ☐   **e** 90 ☐   **f** 68 ☐

2 Divide each number by 10.

  a 50 ☐   **b** 70 ☐   **c** 90 ☐

  d 60 ☐   **e** 120 ☐   **f** 130 ☐

3 Multiply each number by 10.

  a 462 ☐   **b** 2313 ☐   **c** 5456 ☐

4 Divide each number by 10.

  a 450 ☐   **b** 2370 ☐   **c** 12 590 ☐

5 **a** Explain what happens when a number is multiplied by 10.

  _____

  _____

  b Explain what happens when a number is divided by 10.

  _____

  _____

6 Akihiro says: If you multiply a number by 10 you just add a zero.

Do you agree? ☐ Explain.

Show an example.

7 Maddie makes a number. She puts 6 in the ten thousands position, 9 in the thousands position, 7 in the hundreds position, 3 in the tens position and 0 in the ones position. She then divides her number by 10.

Write down her new number.

What is the new value of each number?

6 [   ]   9 [   ]   7 [   ]   3 [   ]

8 Hope is thinking of a 3-digit multiple of 10. She halves it and subtracts 100. She then divides it by 10. Her new number is 26.

What number did she start with?

Show how you worked this out.

9 Amos is thinking of a 2-digit number. He doubles it and adds 100. He then multiplies it by 10. His new number is 2080.

What number did he start with?

Show how you worked this out.

Date: _____   ☺ 😐 ☹

Number

Lesson 2: **Multiplying and dividing by 100**

• Use place value to multiply and divide whole numbers by 100

1 Multiply each number by 100.

a 6 ☐   **b** 2 ☐   **c** 3 ☐

d 9 ☐   **e** 4 ☐   **f** 7 ☐

2 Divide each number by 100.

a 100 ☐   **b** 300 ☐   **c** 700 ☐

d 200 ☐   **e** 600 ☐   **f** 800 ☐

3 Multiply each number by 100.

a 568 ☐   **b** 746 ☐   **c** 8456 ☐

4 Divide each number by 100.

a 5600 ☐   **b** 12 300 ☐   **c** 456 700 ☐

5 **a** Explain what happens when a number is multiplied by 100.

_____

_____

b Explain what happens when a number is divided by 100.

_____

_____

6 Tooka says: If you multiply a number by 100 you just add two zeros.

Do you agree? ☐ Explain.

Show an example.

7 Pia multiplies 458 by 100. This is her product: 45 800

She thinks that her new number is 458 and two zeros. She thinks the 4 is still 4 hundreds, the 5 is still 5 tens and the 8 is still 8 ones.

Is she correct? ☐ Explain your thinking.

_____

_____

8 Ezra is thinking of a 3-digit number. He doubles it and adds 50. He then multiplies it by 10. His new number is 3340.

What number did he start with? ☐

Show how you worked this out.

9 Martha is thinking of a 5-digit multiple of 100. She subtracts 400. She then divides it by 100. Her new number is 432.

What number did she start with? ☐

Show how you worked this out.

Date: _____

Lesson 3: **Rounding to the nearest 10, 100 and 1000**

Number

• Round numbers to the nearest 10, 100 and 1000

1 Round each number to the nearest 10.

    a 46 ⬚   **b** 98 ⬚   **c** 74 ⬚

    d 81 ⬚   **e** 245 ⬚   **f** 632 ⬚

2 Sophie says that 21, 22, 23 and 24 are the only numbers that round to 20.

Which numbers has she missed?

_____

3 Round each number to the nearest 100.

    a 279 ⬚   **b** 435 ⬚   **c** 648 ⬚

    d 4672 ⬚   **e** 6579 ⬚   **f** 8234 ⬚

4 Round each number to the nearest 1000.

    a 3561 ⬚   **b** 7824 ⬚   **c** 6478 ⬚

    d 9054 ⬚   **e** 9773 ⬚   **f** 54 892 ⬚

5 Yakoob thinks you can't round 2098 to the nearest 100, because there is a zero in the hundreds position.

Do you agree? ⬚ Explain your thinking.

_____

_____

6
1 Joseph thinks of a number. He rounds it to the nearest 10. The result is 460.

What could his number be?

Write down all the possibilities.

7
4 Rounding is a useful skill.

Give two reasons why.

_____

_____

_____

8 Jodi estimates the sum of a calculation by rounding the two numbers to the nearest 10 and then adding them.

Her two rounded numbers are 520 and 270.

Her estimate is 790.

What could the original calculation be?

Write down five possibilities.

Number

Lesson 4: **Rounding to the nearest 10 000 and 100 000**

• Round numbers to the nearest 10 000 and 100 000

1 Round each number to the nearest 100.

a 723 ⬚   **b** 345 ⬚   **c** 678 ⬚

d 761 ⬚   **e** 399 ⬚   **f** 511 ⬚

2 Round each number to the nearest 1000.

a 2145 ⬚   **b** 1823 ⬚   **c** 3468 ⬚

d 1528 ⬚   **e** 1291 ⬚   **f** 7953 ⬚

3 Round each number to the nearest 10 000.

a 24 145 ⬚   **b** 18 723 ⬚   **c** 36 468 ⬚

d 15 328 ⬚   **e** 12 891 ⬚   **f** 71 953 ⬚

4 Marie starts with 1 435 861.

She rounds it to the nearest 10 000.

She ends up with 1 445 861.

Is Marie correct? ⬚ Explain your thinking.

_____

_____

5 Amos needs to add 31 423 and 39 494. He thinks a good estimate is 60 000.

Do you agree? ⬚ Explain why.

_____

_____

6 Amita rounds a number. Her result is 1 800 000.

What has she rounded to?

Give two examples of what her number could be.

Give two examples of what her number could **not** be.

7 Henri thinks these numbers would round to 324 670:

324 673, 324 675, 324 668, 324 667.

Spot his mistake.

Why is it a mistake?

_____

_____

8 Adam rounds a number to the nearest 10.

His new number is 47 250.

Write down all the numbers he could have started with.

9 Write instructions on how to round to the nearest 10 000 and 100 000 for someone who doesn't know about rounding.

_____

_____

_____

_____

_____

Date: _____   ☺ ☺ ☹

Lesson 1: **Understanding fractions**

- Understand that the larger the denominator, the smaller the fraction

Number

1 What fraction of each bar is shaded?

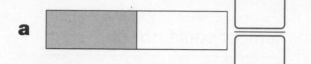

a

b

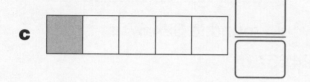

c

d

e Look at the fractions above. Which is the smallest fraction?

2 Shade these fractions on the bars.

a $\frac{1}{3}$

b $\frac{1}{6}$

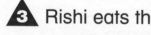

c $\frac{1}{9}$

d Which is the greatest fraction?

3 Rishi eats this fraction of his pizza.

a What fraction does he eat?

b What fraction is left?

4 Tia has a pizza that is the same size as Rishi's.

She eats this fraction.

a What fraction has she eaten?

b What fraction is left?

5 Who has eaten more pizza, Rishi or Tia?

Explain how you know.

_____

_____

6 Barney writes these fractions: $\frac{1}{2}$ $\frac{1}{6}$ $\frac{1}{8}$

Barney says: I think $\frac{1}{8}$ is the greatest fraction.

Is this sometimes, always or never true?

Why?

_____

_____

7 Barney also writes these fractions: $\frac{5}{9}$ $\frac{5}{7}$ $\frac{5}{6}$

4

Barney says: I think $\frac{5}{6}$ is the smallest fraction as it has the smallest denominator.

Is Barney right?   Explain your thinking.

_____

_____

Date: _____

Number

Lesson 2: **Making one whole**

• Understand that fractions can be combined to make one whole

1 Complete these statements.

a $\dfrac{\boxed{}}{2}$ = one whole

b $\dfrac{\boxed{}}{3}$ = one whole

c $\dfrac{\boxed{}}{4}$ = one whole

d $\dfrac{\boxed{}}{5}$ = one whole

e $\dfrac{\boxed{}}{6}$ = one whole

f $\dfrac{\boxed{}}{7}$ = one whole

g $\dfrac{\boxed{}}{8}$ = one whole

h $\dfrac{\boxed{}}{9}$ = one whole

2 Sam gives the answer $\frac{10}{10}$. What question could he have been asked?

_____

3 Moto writes two fractions: $\frac{4}{8}$ $\frac{1}{2}$

a What is the same about the fractions?

_____

b What is different?

_____

c If Moto puts the two fractions together what will she make?

_____

4 Toby says: $\frac{1}{3}$ and $\frac{3}{6}$ together are equivalent to one whole.

Maddie says: That's not right. $\frac{1}{3}$ and $\frac{4}{6}$ together are equivalent to one whole.

Who do you agree with? $\boxed{}$

120

Explain why.

_____

_____

5 Three different pizzas are each cut into 5 slices.

3 slices of the cheese and tomato pizza, 4 slices of the spicy pizza and 2 slices of the vegetable pizza are eaten.

a What fraction of each pizza is **not** eaten?

cheese and tomato = □/□   spicy = □/□   vegetable = □/□

b Explain how you worked out the fractions that were **not** eaten.

_____

_____

6 Complete these statements.

a $\frac{2}{3}$ and $\frac{\square}{3}$ together are equivalent to one whole.

b $\frac{1}{3}$ and $\frac{\square}{3}$ together are equivalent to one whole.

c $\frac{2}{8}$ and $\frac{\square}{8}$ together are equivalent to one whole.

d $\frac{4}{12}$ and $\frac{\square}{12}$ together are equivalent to one whole.

e $\frac{5}{6}$ and $\frac{\square}{6}$ together are equivalent to one whole.

f $\frac{2}{\square}$ and $\frac{3}{\square}$ together are equivalent to one whole.

Date: _____

Lesson 3: **Equivalent fractions**

• Recognise that two proper fractions can be equivalent in value

1 Shade half of each shape.

a
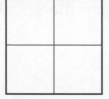
How many quarters
have you shaded?

b
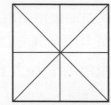
How many eighths
have you shaded?

c

How many halves
have you shaded?

d
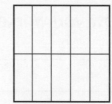
How many tenths
have you shaded?

2 These shapes show equivalent fractions. What are they?

a
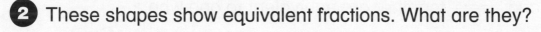

b

c

d
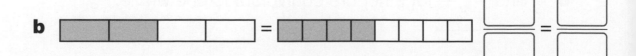

Number

3 Write 'True' or 'False' for each of these statements.

a Two-quarters is equivalent to one-half. _____

b $\frac{4}{8} = \frac{1}{4}$ _____

c One out of five is the same as five out of 10. _____

d $\frac{1}{2} = \frac{5}{10}$ _____

e Three-fifths is equivalent to six-tenths. _____

f $\frac{2}{4} = \frac{2}{8}$ _____

g four-sixths is equal to two-thirds _____

4 This strip shows $\frac{1}{5}$ shaded.

Leah thinks that if she halves each fifth she will make tenths.

Her $\frac{1}{5}$ would then be $\frac{2}{10}$. Do you agree? ☐ Explain why.

_____

_____

_____

5 Sort these fractions into two groups.

 $\frac{2}{3}$ $\frac{4}{5}$ $\frac{6}{9}$ $\frac{20}{25}$ $\frac{12}{15}$ $\frac{12}{18}$

Group 1		Group 2	

How did you sort them?

_____

_____

_____

Date: _____

Number

Lesson 4: **Comparing and ordering fractions**

• Use knowledge of equivalence to compare and order proper fractions using the symbols =, > and <

1 Colour each fraction and then write which is greater.

a $\frac{1}{4}$ $\frac{1}{2}$ ☐/☐ is greater than ☐/☐

b $\frac{1}{4}$ $\frac{1}{3}$ ☐/☐ is greater than ☐/☐

c $\frac{1}{2}$ $\frac{1}{5}$ ☐/☐ is greater than ☐/☐

2 Write < or > to show which fraction is smaller or greater.

Example:

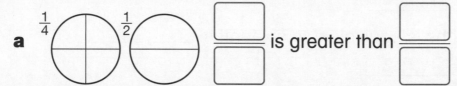

$\frac{2}{4}$ < $\frac{3}{5}$

a ☐/☐ ☐ ☐/☐

b ☐/☐ ☐ ☐/☐

c ☐/☐ ☐ ☐/☐

3 Use the fraction wall in **4** to order each set fractions, smallest first.

a $\frac{1}{4}, \frac{7}{10}, \frac{4}{8}$ ☐ , ☐ , ☐

b $\frac{1}{10}, \frac{4}{5}, \frac{1}{2}$ ☐ , ☐ , ☐

c $\frac{5}{8}, \frac{5}{10}, \frac{1}{3}$ ☐ , ☐ , ☐

d $\frac{1}{5}, \frac{8}{10}, \frac{4}{8}$ ☐ , ☐ , ☐

4 Use the fraction wall to order each set of fractions, smallest first.

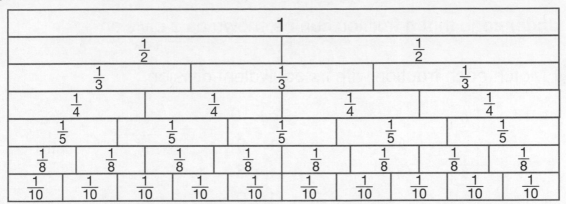

a $\frac{2}{3}$, $\frac{2}{4}$, $\frac{1}{2}$, $\frac{4}{10}$ ⬚ , ⬚ , ⬚ , ⬚   **b** $\frac{9}{10}$, $\frac{3}{5}$, $\frac{7}{8}$, $\frac{2}{4}$ ⬚ , ⬚ , ⬚ , ⬚

c $\frac{2}{8}$, $\frac{2}{5}$, $\frac{1}{10}$, $\frac{2}{3}$ ⬚ , ⬚ , ⬚   **d** $\frac{4}{8}$, $\frac{6}{10}$, $\frac{4}{5}$, $\frac{1}{4}$ ⬚ , ⬚ , ⬚ , ⬚

5 Make up five different fractions from the fraction wall in **4**.
They must all have a numerator greater than 1.
Order them from smallest to greatest. Do this twice.

⬚ , ⬚ , ⬚ , ⬚ , ⬚   ⬚ , ⬚ , ⬚ , ⬚ , ⬚

Explain what you did.

_____

_____

6 Ibrahim likes sweets. Should he choose $\frac{1}{5}$ ⬚
of a bag of sweets or $\frac{1}{3}$ of a bag of sweets?

Explain why. _____

_____

7 Write your own fraction comparison question.

_____

_____

Date: _____

Number

Lesson 1: **Fractions as division**

- Understand that a fraction can be shown as a division

1 Match each fraction with its equivalent division.

$\frac{1}{6}$

$1 \div 10$
$1 \div 8$
$1 \div 5$
$1 \div 9$
$1 \div 6$

$\frac{1}{10}$

$\frac{1}{8}$

$\frac{1}{5}$

$\frac{1}{9}$

2 Sonja says that $\frac{1}{2}$ is the same as $2 \div 1$. Do you agree? ⬜ Why?

4

_____

_____

3 Write the division and the fraction that matches each bar.

a

b

c

d

4 Maemi, Umeki, Eiko and Uta share 2 bananas equally.

a What fraction do they cut each banana into?

b What fraction of a whole banana do each of them eat? ⬜

Number

5 Bobby shares his birthday cake between himself and his 7 friends. He thinks everyone will get $\frac{1}{7}$ of his cake.

Do you agree? ☐ Why?

_____

_____

6 Carly has a bar of chocolate. She divides it into 5 parts.

a Write the division statement to show this.

☐

b What fraction is each part?

☐
☐

c Carly eats 3 parts. What fraction does she eat?

☐
☐

d What fraction is left?

☐
☐

7 This table shows the number of sweets in different packets. Fill in the missing information.

Number in the packet	Fraction eaten	Division	Number eaten
24	$\frac{1}{8}$		
32		$1 \div 4$	
40		$1 \div 5$	
55	$\frac{1}{5}$		
56		$1 \div 7$	
72		$1 \div 9$	
84	$\frac{1}{12}$		
100	$\frac{1}{10}$		

Date: _____

Number

Lesson 2: **Fractions of quantities**

• Understand that fractions can act as operators

 Colour each shape to show the fraction.

a

$\frac{3}{4}$

b

$\frac{1}{2}$

c

$\frac{1}{4}$

d

$\frac{1}{10}$

2 Each of the shapes in **1** has a value of 20.

What is the value of each part you have shaded?

a [　]   **b** [　]   **c** [　]   **d** [　]

3 Find the value of each of these. Then write what else you know.

a $\frac{1}{3}$ of 21 [　] [　]   **b** $\frac{1}{4}$ of 24 [　] [　]

c $\frac{1}{5}$ of 30 [　] [　]   **d** $\frac{1}{6}$ of 42 [　] [　]

e $\frac{1}{7}$ of 56 [　] [　]   **f** $\frac{1}{7}$ of 63 [　] [　]

4 Kwame cuts three strips of paper into different lengths.

a The first strip is 48 cm long. He cuts it into eighths.

How long will each eighth be? [　] cm

b His second strip of paper is a metre (100 cm) long. He cuts it into quarters.

How long will each quarter be? ☐ cm

c The last strip is 70 cm long. He cuts it into tenths.

How long will each tenth be? ☐ cm

 What fraction of each shape is shaded?

a Write your answer in sixths: ☐/☐

b Write your answer in quarters: ☐/☐

c Write your answer in eighths

and quarters: ☐/☐ or ☐/☐

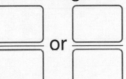

d Write your answer in tenths and fifths:

☐/☐ or ☐/☐

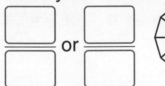

6 Write two fractions of number calculations that have the answer shown. Example: 26

$\frac{1}{2}$ of 52 = 26 $\frac{1}{4}$ of 104 = 26

a 21

b 32

Date: _____

Number

Lesson 3: **Adding fractions**

• Add fractions with the same denominator

1 This bar shows fifths.

Use it to help you add these .

a $\frac{1}{5} + \frac{2}{5} = \dfrac{\boxed{}}{\boxed{}}$　　**b** $\frac{1}{5} + \frac{3}{5} = \dfrac{\boxed{}}{\boxed{}}$　　**c** $\frac{1}{5} + \frac{4}{5} = \dfrac{\boxed{}}{\boxed{}}$　　**d** $\frac{2}{5} + \frac{3}{5} = \dfrac{\boxed{}}{\boxed{}}$

2 Toby says:

> Adding fractions is the same as adding whole numbers.

Do you agree? $\boxed{}$ Why? Give an example.

_____

_____

3 Add these fractions.

a $\frac{3}{7} + \frac{2}{7} = \dfrac{\boxed{}}{\boxed{}}$　　　**b** $\frac{1}{8} + \frac{2}{8} = \dfrac{\boxed{}}{\boxed{}}$　　　**c** $\frac{3}{5} + \frac{1}{5} = \dfrac{\boxed{}}{\boxed{}}$

d $\frac{2}{9} + \frac{5}{9} = \dfrac{\boxed{}}{\boxed{}}$　　　**e** $\frac{1}{3} + \frac{1}{3} = \dfrac{\boxed{}}{\boxed{}}$　　　**f** $\frac{1}{6} + \frac{5}{6} = \dfrac{\boxed{}}{\boxed{}}$

4 Demi says:

> I can't add $\frac{3}{5}$ and $\frac{4}{5}$ because 3 add 4 is 7, which is more than 5.

Do you agree? $\boxed{}$ Why?

_____

_____

5 Fill in the missing fractions to make the addition calculations correct.

a $\frac{1}{3} + \dfrac{\boxed{}}{\boxed{}} = \frac{3}{3}$ **b** $\frac{2}{5} + \dfrac{\boxed{}}{\boxed{}} = \frac{4}{5}$ **c** $\dfrac{\boxed{}}{\boxed{}} + \frac{5}{6} = \frac{10}{6}$

 6 Adam says:

> If I know that $\frac{5}{9} + \frac{3}{9} = \frac{8}{9}$, I also know that $\frac{3}{9} + \frac{5}{9} = \frac{8}{9}$.

Do you agree? $\boxed{}$

Why?

_____

_____

7 **a** Turn these fractions into equivalent fractions that all have the same denominator. Show your working out in the box below.

$\frac{3}{4} = \dfrac{\boxed{}}{\boxed{}}$ $\frac{1}{2} = \dfrac{\boxed{}}{\boxed{}}$ $\frac{5}{8} = \dfrac{\boxed{}}{\boxed{}}$

Now add them all together. What is the total? $\dfrac{\boxed{}}{\boxed{}}$

b Do the same for these.

$\frac{1}{4} = \dfrac{\boxed{}}{\boxed{}}$ $\frac{1}{2} = \dfrac{\boxed{}}{\boxed{}}$ $\frac{3}{8} = \dfrac{\boxed{}}{\boxed{}}$ $\frac{5}{16} = \dfrac{\boxed{}}{\boxed{}}$

Now add them all together. What is the total? $\dfrac{\boxed{}}{\boxed{}}$

Date: _____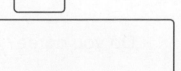

Number

Lesson 4: **Subtracting fractions**

• Subtract fractions with the same denominator

1 This bar shows sixths.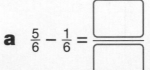
Use it to help you subtract these fractions.

a $\frac{5}{6} - \frac{1}{6} = \boxed{}$ **b** $\frac{5}{6} - \frac{2}{6} = \boxed{}$ **c** $\frac{5}{6} - \frac{3}{6} = \boxed{}$ **d** $\frac{5}{6} - \frac{4}{6} = \boxed{}$

2 Freddie says:

> Subtracting fractions is the same as subtracting whole numbers.

Do you agree? $\boxed{}$ Why? Give an example.

_____

_____

3 Subtract these fractions.

a $\frac{4}{7} - \frac{2}{7} = \boxed{}$ **b** $\frac{6}{8} - \frac{2}{8} = \boxed{}$ **c** $\frac{5}{5} - \frac{1}{5} = \boxed{}$

d $\frac{7}{9} - \frac{5}{9} = \boxed{}$ **e** $\frac{2}{3} - \frac{1}{3} = \boxed{}$ **f** $\frac{4}{6} - \frac{1}{6} = \boxed{}$

4 Adnan says:

> I can't subtract $\frac{4}{5}$ from $\frac{8}{5}$ because $\frac{8}{5}$ is greater than one whole.

Do you agree? $\boxed{}$ Why?

_____

_____

Number

5 Fill in the missing fractions to make the subtraction calculations correct.

a $\frac{7}{8} - \dfrac{\square}{\square} = \frac{3}{8}$ **b** $\frac{6}{5} - \dfrac{\square}{\square} = \frac{3}{5}$ **c** $\dfrac{\square}{\square} - \frac{5}{6} = \frac{4}{6}$

d $\dfrac{\square}{\square} - \frac{2}{7} = \frac{3}{7}$ **e** $\frac{11}{8} - \dfrac{\square}{\square} = \frac{7}{8}$ **f** $\dfrac{\square}{\square} - \frac{7}{10} = \frac{8}{10}$

6 Beatrice says: If I know that $\frac{7}{12} - \frac{3}{12} = \frac{4}{12}$, I also know that $\frac{7}{12} - \frac{4}{12} = \frac{3}{12}$.

Do you agree? $\square$

Why?

_____

_____

7 a Turn these fractions into equivalent fractions that all have the same denominator.

$\frac{3}{6} = \dfrac{\square}{\square}$ $\frac{2}{3} = \dfrac{\square}{\square}$ $\frac{5}{12} = \dfrac{\square}{\square}$

b Use pairs of these fractions to make three subtraction calculations.

8 For each subtraction calculation that you made in **7**, write the other three facts that you know.

Date: _____

133

Number

Lesson 1: **What is a percentage? (1)**

* Understand that a percentage is part of a whole

1 Write three different examples of where you might see percentages.

_____

_____

_____

2 Draw a ring around the percentages.

34    $\frac{1}{2}$ 54   80% $\frac{1}{5}$ 25% $\frac{2}{3}$ 20%   18

How do you know that they are percentages?

_____

_____

3 Shade these shapes to show 100%.

4 Adil says:

To show one whole as a percentage you write $\frac{1}{100}$.

Do you agree? ☐

Explain why.

_____

_____

Number

5 Jamie thinks that if one whole is 100%, then $\frac{1}{2}$ a whole is 50%.

Do you agree? [ ] Explain why.

_____

_____

6 Draw a ring around the odd one out.

20%   40%   60% $\frac{1}{3}$ 80%

Why is it the odd one out?

_____

_____

7 Sophie says:

> The battery on my tablet shows 99%.
> I think I will need to charge it up soon.

Do you agree? [ ] Explain why.

_____

_____

8 If you know 100% is equivalent to $\frac{100}{100}$, what else do you know?
Write ten other facts.

_____

Date: _____

Number

Lesson 2: **What is a percentage? (2)**

• Understand that a percentage is part of a whole

1 Draw a ring around the odd one out.

$\frac{2}{5}$ $\frac{4}{5}$ $\frac{1}{5}$ 99% $\frac{3}{5}$

Why is it the odd one out? _____

2 Draw a ring around the highest percentage.

10%   65%   21%   82%   53%

How do you know it is the highest?

 3 Write the missing percentage to make 100%.

a 100% = 10% + ⬚%   **b** 100% = 60% + ⬚

c 100% = 15% + ⬚%   **d** 100% = ⬚ + 99%

e 100% = ⬚ + 2%   **f** 100% = ⬚ + 25%

 4 Mara and Tyra have been shopping in the sales. They both bought a pair of shoes. Before the sale both pairs of shoes cost $50.

Mara says: There was a 40% discount in the shop I went to. I think my shoes cost less than yours.

Tyra says: There was a 60% discount in the shop I went to. I think my shoes were cheaper.

Who do you agree with? ⬚

Explain why.

_____

_____

Number

5 Write these percentages as fractions out of 100.

a 10% ▭/▭ **b** 30% ▭/▭ **c** 55% ▭/▭

6 Adnan says: 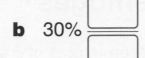 If I know 20% + 80% equals 100%, I know three more percentage facts and four fraction facts.

Prove what Adnan says is true.

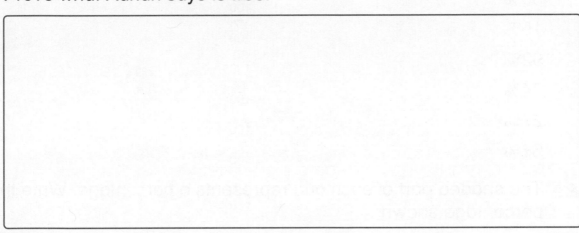

7 Sort these percentages into two groups.

20%   65%   70%   30%

15%   90%   85%   45%

How did you sort them?

_____

How else could you sort them?

_____

Show your two new groups here.

Number

Lesson 3: **Expressing hundredths as percentages**

• Understand that a percentage is part of a whole

1 Draw lines to match each percentage with its equivalent fraction.

20% $\frac{61}{100}$

30% $\frac{20}{100}$

15% $\frac{27}{100}$

92% $\frac{30}{100}$

74% $\frac{74}{100}$

27% $\frac{15}{100}$

61% $\frac{92}{100}$

2 The shaded part of each grid represents a percentage. Write the percentage shown.

a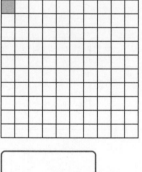

b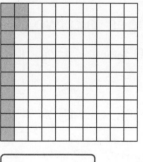

c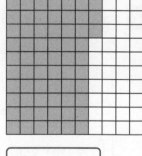

3 Shade the percentages.

a 19%   b 55%   c 73%

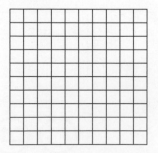   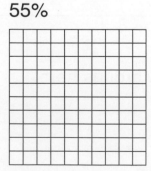   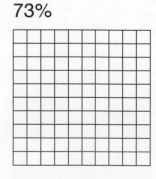

4 Write each of these as a percentage.

a 13 out of 100 cars are grey

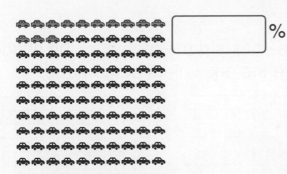 %

b 29 out of 100 cakes are chocolate.

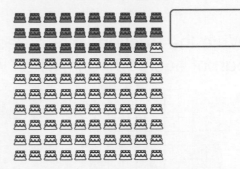 %

c 76 out of 100 days are sunny.

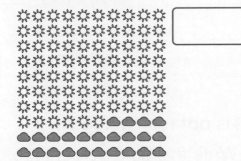 %

d 81 out of 100 people are happy.

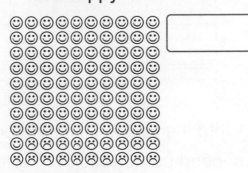 %

5 Look at the pictures in **4**. Write the percentage.

a Cars that are **not** grey

b Cakes that are **not** chocolate

c Days that are **not** sunny

d People that are **not** happy

6 Write your own 'out of 100' statement, similar to those in **4**. Write the percentage you have described.

_____

_____ %

Date: _____

139

Lesson 4: **Expressing fractions as percentages**

Number

• Understand that fractions can be represented as percentages

1 Write the fraction, in hundredths, that is represented by the shaded part of each 100 grid. Then write the equivalent percentage.

a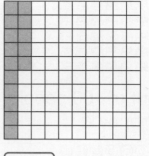

b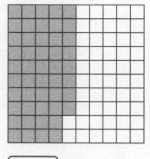

c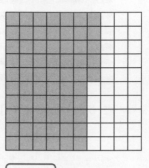

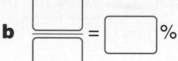 = ▢ %   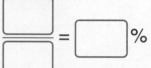 = ▢ %   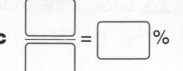 = ▢ %

2 Write the fraction, in hundredths, that is **not** represented by the shaded part of each 100 grid in **1**. Write this as a percentage.

a ▢/▢ = ▢ %   **b** ▢/▢ = ▢ %   **c** ▢/▢ = ▢ %

3 Shade the fraction of the grid shown and write the percentage.

a $\frac{1}{4}$ = ▢ %   **b** $\frac{3}{10}$ = ▢ %   **c** $\frac{75}{100}$ = ▢ %

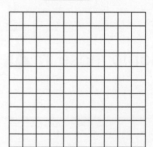   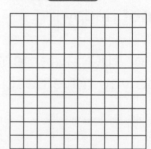   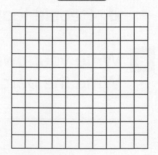

Number

4 Write the fraction and percentage that you did **not** shade in each grid in **3**.

a ⬜/⬜ = ⬜ %   **b** ⬜/⬜ = ⬜ %   **c** ⬜/⬜ = ⬜ %

5 Write each of these as a percentage.

a 3 out of every 4 children in Class 4 are boys. ⬜ %

b 75 out of every 100 cars in the car park are black. ⬜ %

c 7 out of every 10 sweets in a bag are chewy. ⬜ %

d 25 out of every 100 shapes are prisms. ⬜ %

e 15 out of every 100 pets are rabbits. ⬜ %

f 85 out of every 100 coins are cents. ⬜ %

6 Write an equivalent fraction for each percentage.

a 25% = ⬜/⬜   **b** 10% = ⬜/⬜   **c** 20% = ⬜/⬜

d 75% = ⬜/⬜   **e** 50% = ⬜/⬜   **f** 80% = ⬜/⬜

7 Write the percentage equivalent for each fraction.

a $\frac{1}{10}$ = ⬜   **b** $\frac{7}{10}$ = ⬜   **c** $\frac{97}{100}$ = ⬜

d $\frac{1}{2}$ = ⬜   **e** $\frac{1}{4}$ = ⬜   **f** $\frac{75}{100}$ = ⬜

Date: _____

Lesson 1: **Units of time**

> • Understand the relationship between units of time, and convert between them

Geometry and Measure

1 How many minutes are there in these numbers of seconds?

a 120 seconds ☐ minutes

b 180 seconds ☐ minutes

c 240 seconds ☐ minutes

d 1200 seconds ☐ minutes

2 How many days are there in these numbers of hours?

a 24 hours ☐ day   **b** 48 hours ☐ days

c 240 hours ☐ days   **d** 120 hours ☐ days

3 If you know that there are 52 weeks in one year, what else do you know? Write five more facts.

4 Use the words in the box to describe the time intervals.

> minutes   hours   days   weeks   seconds

a How long you watch TV for each week. _____

b How long it takes to get to school in the morning. _____

c How long it is before bedtime. _____

d How long it takes to sharpen a pencil. _____

 5 Stefan says: < There can't be 60 seconds in a minute because there are 60 minutes in an hour.

Do you agree? [   ] Why?

_____

 6 Put a ✗ in the box next to the statement that is incorrect.

120 minutes = 2 hours [   ]   70 days = 10 weeks [   ]

156 weeks = 3 years [   ]   240 seconds = 3 minutes [   ]

Why is it incorrect?

_____

7 Write all the months in the year that have 30 days.

```
[                                                          ]
```

8 Write all the months in the year that have 31 days.

```
[                                                          ]
```

9 'There are 28 days in February.'

Is this always, sometimes or never true?

_____

Why?

_____

Date: _____

143

Lesson 2: **Linking analogue and 12-hour digital times**

Geometry and Measure

• Read and record time accurately on analogue and 12-hour digital clocks

1 Write these analogue times as digital times.

a 25 minutes past 3 [:]　　**b** 10 minutes past 4 [:]

c 35 minutes past 6 [:]　　**d** 45 minutes past 8 [:]

e 55 minutes past 10 [:]　　**f** 15 minutes past 7 [:]

2 Order these times, from earliest to latest.

| 4:30 | 4:45 | 4:05 | 4:50 | 4:20 | 4:55 |

_____

 Write an equivalent analogue time beside each digital time.

a 10:46 _____

b 4:53 _____

c 6:11 _____

 Write the time on each clock in two different ways.

a

b

c

[ ]　　　[ ]　　　[ ]

[:]　　　[:]　　　[:]

Geometry and Measure

5 Show or write the missing times in each question.

a

10:22 | a.m.● p.m.○

b

07:33 | a.m.● p.m.○

c

12 minutes to 8 in the evening

| : | a.m.○ p.m.○

6 Look at the clocks in **5**. What is the time interval, in minutes, between clocks **b** and **c**?

| | minutes. Show how you worked this out.

7 These clock faces have lost their numbers! Estimate the times that they are showing. Write each time in two different ways.

a

:

b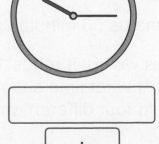

:

c

:

Date: _____

Lesson 3: **Linking 12- and 24-hour time**

Geometry and Measure

> • Read and record time accurately on analogue and 12- and 24-hour digital clocks

1 Draw lines to match the times.

8:15 p.m.	07:15
5:15 p.m.	13:15
10:15 a.m.	15:15
3:15 p.m.	10:15
1:15 p.m.	20:15
7:15 a.m.	17:15

2 Order these times, from earliest to latest.

> 06:34   22:16   19:45   13:56   01:15

_____

3 Write 'True' or 'False' for each statement.

a 13:25 is the same as 1:25 a.m. _____

b 24 minutes past 6 in the evening is the same as 18:24. _____

c 22:10 is the same as 10:10 in the evening. _____

d 14 minutes past 11 in the morning is the same as 11:14 a.m. _____

e 16:56 is the same as 56 minutes past 4 in the morning. _____

4 Reuben looks at his watch. It reads 13:25. He is 20 minutes late to meet his friend. At what time should he have been there? Write your answer in four different ways.

Geometry and Measure

5 Ruth goes for a bike ride. She leaves home at 15:05.
She gets back at 15:48.
How long is she out on her bike?
Use the time line to help you.

05 10 15 20 25 30 35 40 45 50 55   05 10 15 20 25 30
15:00   16:00

6 Ben goes out with his friend. He leaves home at 09:02.
He gets back at 09:56.
How long is he away from home?
Use the time line to help you.

05 10 15 20 25 30 35 40 45 50 55   05 10 15 20 25 30
09:00   10:00

7 Sort these times into two groups.

> 13 minutes past 3 in the afternoon   21:17   5:30 p.m.
> 25 minutes to 8 in the morning   14:25   07:38
> 52 minutes past 12 in the afternoon   15:16

Group 1	Group 2

How did you sort them?

_____

_____

Is there another way to sort them?

_____

_____

Date: _____

147

Geometry and Measure

Lesson 4: **Timetables**

• Interpret and use timetables in 12- and 24-hour times

You will need
• bus and plane timetables

1 Calculate these time differences. Draw a time line to help you.

a 7:30 to 7:55

b 11:00 to 11:45

c 12:09 to 12:40

2 Calculate these time differences. Draw a time line to help you.

a 8:25 a.m. to 8:50 a.m.

b 11:05 a.m. to 11:56 p.m.

c 18:04 to 18:38

d 14:05 to 14:55

3 Use the bus timetable to answer the questions.

a Where is Bus D at 1:22 p.m.? _____

b Which bus is at Brook's Hill at 9:40 a.m.? _____

c Which stop is before Luton Street? _____

d How long does it take Bus E to get from the library to the railway station? _____

e Write one statement about the information in the timetable.

_____

4 Use the plane timetable to answer the questions.

a How long is the 09:15 flight to Germany?

b What time does the flight leave the UK that arrives in Germany at 22:50?

c How long is the 15:06 flight to Germany?

d Bertie has to get to the airport 2 hours before his 22:38 flight to Germany.

What time does he need to get to the airport?

e The 17:05 flight to Germany arrived 1 hour late.

What time did it arrive in Germany?

5 Fill in a timetable showing imaginary flight times from one country to another. Make up your own times. Write three questions that you can ask from your timetable.

Depart	Arrive

_____

_____

_____

Date: _____

Geometry and Measure

Lesson 1: **Combining polygons**

• Investigate shapes made by combining two or more shapes

1 Name these polygons.

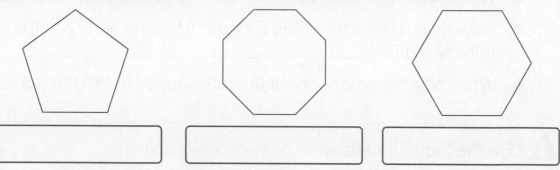

2 Why are the shapes in **1** regular?

◄4

_____

_____

_____

3 Here are three octagons.

◄5

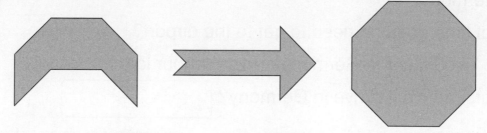

a Write two ways they are alike.

_____

_____

b Write two ways they are different.

_____

_____

4 Mandisa says:

Sketch Mandisa's shape.

> My shape is an irregular pentagon. It has three right angles and two pairs of sides that are the same length.

5 Trey puts these two regular triangles together.

He makes a 4-sided shape. Sketch Trey's shape.

6 Sketch a hexagon in each section of this Carroll diagram.

	Right angles	**No right angles**
Symmetrical		
Not symmetrical		

Date: _____

Geometry and Measure

Lesson 2: **Combining 4-sided shapes**

• Combine two or more 4-sided shapes

1 Name these shapes.

_____   _____

2 Explain why a square is a rectangle.

⟨5⟩

_____

_____

3 Sketch an irregular 4-sided shape.

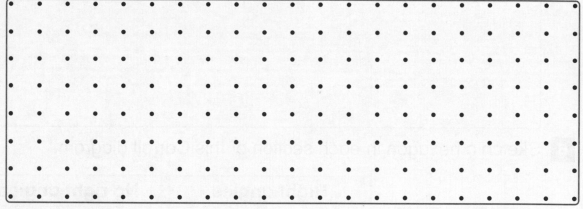

Write the properties of your shape, stating why it is irregular.

_____

_____

4 Sketch two different rectangles.

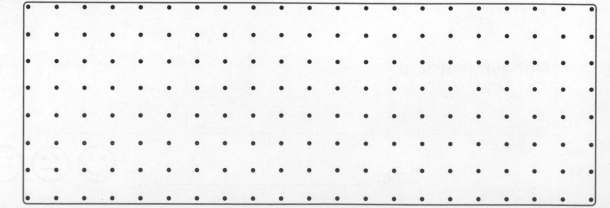

Geometry and Measure

Describe their properties.

_____

_____

5 Sketch two identical 4-sided shapes side by side. The side of one shape must touch one side of the other shape.

How many sides does your new shape have? ☐

6 Create a tessellating pattern by combining two different 4-sided shapes.

Describe your pattern.

_____

_____

_____

Date: _____   ☺ ☺ ☹

153

Geometry and Measure

Lesson 3: **Symmetry of 2D shapes**

• Identify horizontal, vertical and diagonal lines of symmetry

1 Draw all the lines of symmetry on these shapes.

2 Inside each shape write the number of lines of symmetry it has.

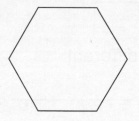

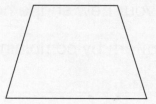

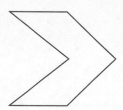

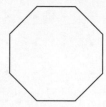

3 Explain why this shape is not symmetrical.

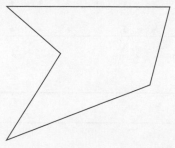

_____

_____

4 Draw two different 4-sided shapes that each have one line of symmetry.

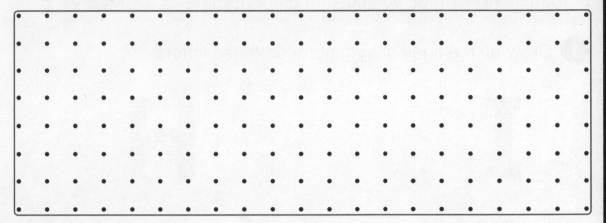

5 Draw a shape made up from six squares that has two lines of symmetry.

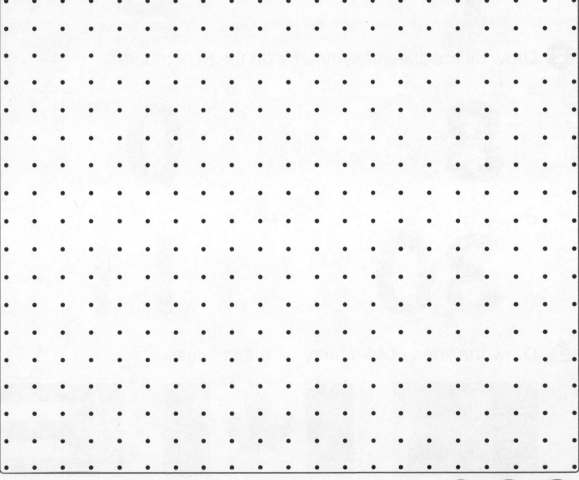

Geometry and Measure

Date: _____

Lesson 4: **Symmetry in real life**

• Identify horizontal, vertical and diagonal lines of symmetry

1 Draw all the lines of symmetry on these letters.

a

I

b

H

c

T

d

B

2 Draw all the lines of symmetry on these numbers.

a

8

b

0

c

30

d

II

3 Draw the lines of symmetry on these flags.

4 These are halves of some shapes. Draw the whole shapes.
What shapes have you drawn? Label each one.
How many lines of symmetry does each shape have?

a
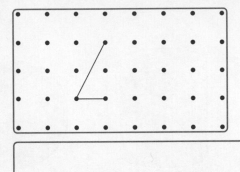

b
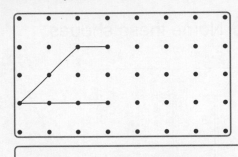

c
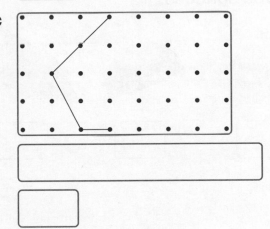

d
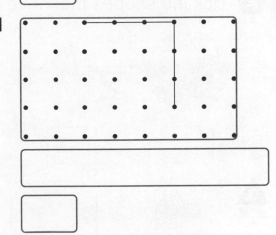

5 Design a pattern on the squares. Your pattern needs to have at least two diagonal lines of symmetry.

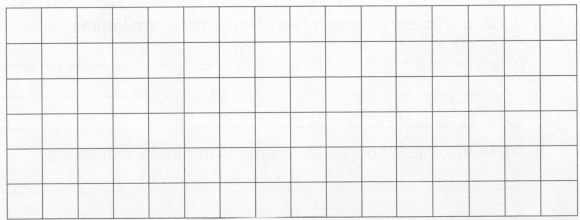

Date: _____

157

Lesson 1: **Shapes with curved surfaces**

- Identify 2D faces of 3D shapes and describe their properties

Geometry and Measure

1 Name these shapes.

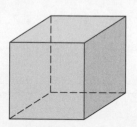   

_____ _____ _____ _____

2 Tick the shapes that have curved surfaces.

   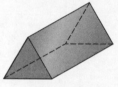

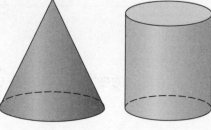

a How are these shapes alike? Write three similarities.

_____

_____

b How are these shapes different? Write three differences.

_____

_____

Geometry and Measure

4 List the properties of these shapes.

   a Cylinder

   _____

   _____

   b Cone

   _____

   _____

5 Portia says:

> I think a cone has two faces.

Do you agree? ⬚

Explain.

_____

_____

6 I have one curved and two circular faces, two circular edges and I can only roll in one direction. What shape am I?

_____

Make up your own clues like this for a cone.

_____

_____

7 Abiole says:

> A cuboid has 6 square faces, 12 edges and 8 vertices.

Is this always, sometimes or never true? ⬚⬚⬚⬚⬚

Explain why.

_____

_____

Date: _____

Geometry and Measure

Lesson 2: **Prisms**

• Identify different prisms and describe their properties

1 Name these shapes.

a 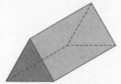 **b** 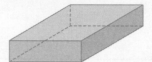 **c**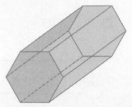

_____   _____   _____

2 How can you identify a prism?

_____

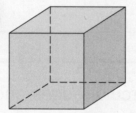 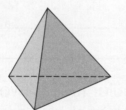 

Which is the odd one out? [   ] Explain why.

_____

_____

4 Ruby thinks the first shape is a cube. She doesn't think the second shape can be a cube because it looks different.

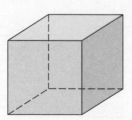 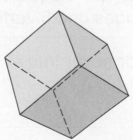

Do you agree? [   ] Explain.

_____

5 **a** How are these shapes the same?

Write three similarities.

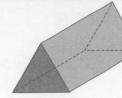 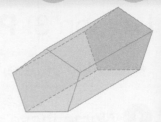

_____

_____

_____

b How are these shapes different?

Write three differences.

_____

_____

_____

6 List the properties of these shapes.

a Pentagonal prism

_____

_____

b Triangular prism

_____

_____

7 Prisms are used in many aspects of everyday life such as in food packaging. Think of three more examples where prism shapes are commonly used. Write about how they are used and why you think that particular prism is used.

_____

_____

_____

_____

Date: _____

Lesson 3: **Pyramids**

• Identify different pyramids and describe their properties

1 Name these shapes.

a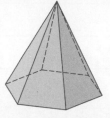

b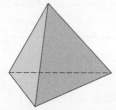

c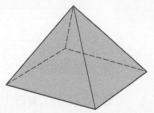

_____ _____ _____

2 How can you identify a pyramid?

_____

3 Jodie thinks the first shape is a square-based pyramid. She thinks the second cannot be a square-based pyramid because it looks different.

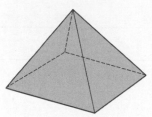 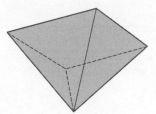

Do you agree? [   ]

Explain.

_____

_____

4 Look at these two shapes.

a How are these shapes the same?

Write three similarities.

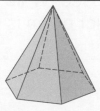 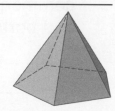

_____

_____

Geometry and Measure

b How are these shapes different?

Write three differences.

_____

_____

_____

5 List the properties of these shapes.

a Triangular-based pyramid

_____

_____

b Square-based pyramid

_____

_____

6 For each statement, circle whether it is sometimes, always or never true. Explain your answers, using drawings if you wish.

a A pyramid has one vertex.

sometimes   always   never

b Shapes with curved surfaces are not pyramids.

sometimes   always   never

c Shapes with flat faces have faces of different shapes.

sometimes   always   never

Date: _____

Geometry and Measure

Lesson 4: **Nets**

- Match nets to their corresponding 3D shapes

1 Is this the net of a cube?

4 Explain your answer

_____

_____

2 What is the net of a shape?

_____

_____

3 Draw lines to match each net to its shape.

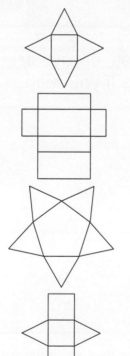

triangular prism

square-based pyramid

cuboid

pentagonal pyramid

4 How is the net of a cuboid the same as the net of a cube?

5 _____

How is it different?

_____

5 **a** Draw a ring around the net of a tetrahedron (triangular-based pyramid).

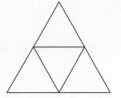

b Explain how you know.

_____

_____

6 Rami says:

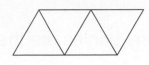

I've drawn a net for a cube.

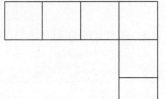

Do you agree? [ ] Explain.

_____

_____

7 Beth says:

It is easy to draw the net of a pyramid. All you do is draw the shape of the name of the pyramid. Then you add triangles to each side of that shape.

Do you agree? [   ]

Explain why. Include a sketch to show your thinking.

_____

_____

_____

Date: _____

Lesson 1: **Angles all around us**

Geometry and Measure

• Recognise angles

1 How many angles are there in each shape?

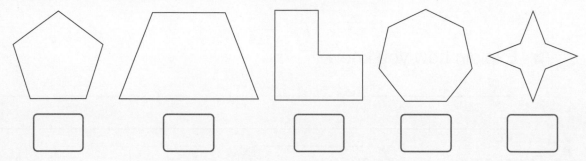

2 Tick the shapes that have at least one right angle.

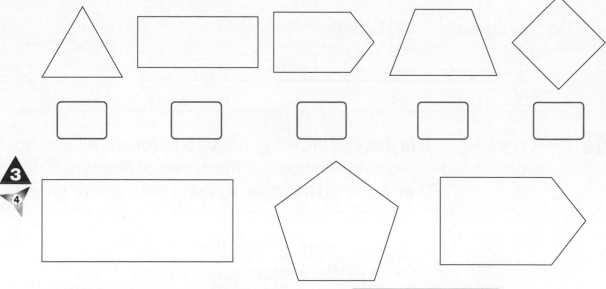

a Which shape is the odd one out?
Why?

_____

b Which other shape could be the odd one out?
Why?

_____

4 Solomon says:

> I know that an 8-sided shape has 8 angles.

Do you agree? ☐

Why?

_____

5 Look at this shape.

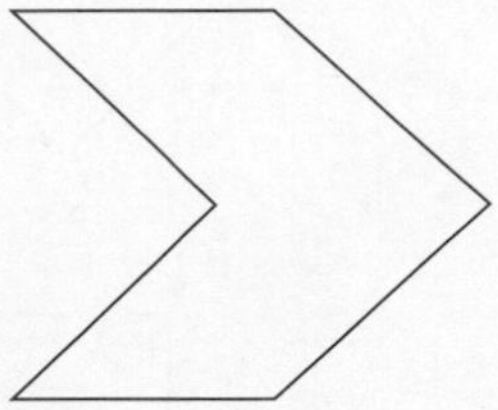

Write down three things that you notice about the angles.

_____

_____

_____

6 I have four sides. They are not all equal. I have four right angles.

What shape am I? _____

Make up your own clues like this for an irregular pentagon.

_____

_____

7 Sophie says:

> My shape has 2 right angles, 2 angles that are greater than a right angle and 2 angles that are smaller than a right angle. What is my shape?

Alex says:

> It's a pentagon.

Do you agree? ☐ Explain why.

_____

Date: _____ ☺ ☹ ☹

Lesson 2: **Right angles**

• Identify and combine right angles

1 Tick the right angles.

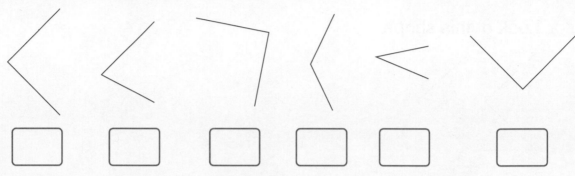

☐   ☐   ☐   ☐   ☐   ☐

2 **a** How many right angles make a straight line? ☐

    b Draw an angle on a straight line.

3 **a** How many degrees are there in a whole turn? ☐

    b How many right angles is this equivalent to? ☐

4 **a** Draw a diagram to show 270° made from right angles.

    b How many degrees are there in a three-quarter turn? ☐

    c How many right angles is this equivalent to? ☐

Geometry and Measure

5 Tick the pair of right angles that shows 180°.

A   B   C   D

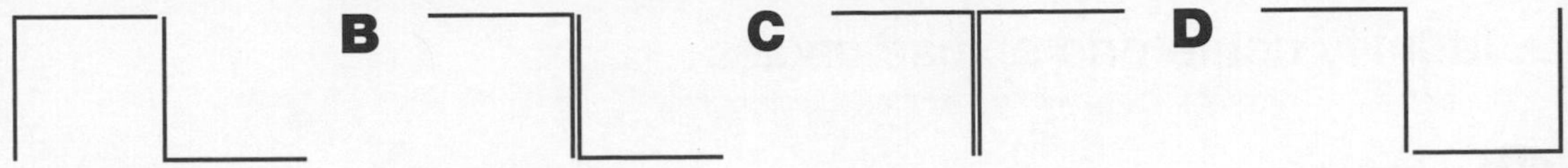

Explain why.

_____

_____

6 Kodi says: A right angle is made when a vertical line and a horizontal line meet.

Is this sometimes, always or never true? [ ]
Explain why.

_____

_____

7 Patsy says: If I cut all the corners off this square, I can make a whole turn of 360°.

Is Patsy correct? [ ] Draw a diagram in the space below to prove it.

Lesson 3: **Acute and obtuse angles**

• Identify acute and obtuse angles

1 Label these angles as acute, right or obtuse.

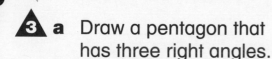

_____ _____ _____ _____

2 Write a definition for a right angle.

⑤ _____

3 **a** Draw a pentagon that has three right angles.

   b What can you say about the other two angles?

_____

_____

4 **a** Draw a 4-sided shape with two right angles.

   b Describe the other two angles.

_____

_____

5 Tom says: An acute angle is small so it is very cute.

⑦

Is this a good definition for an acute angle? ☐ Explain.

_____

____ _____

Geometry and Measure

Geometry and Measure

6 Write a definition for an obtuse angle.

_____

_____

7 Draw a shape with a right angle, two acute angles and an obtuse angle inside this circle. Make sure the sides of your angles touch the outside of the circle.

Label each angle.

8 Perdita thinks she can draw a triangle with one right angle, one acute angle and one obtuse angle. Write an explanation with words and diagrams to show why this is impossible.

Date: _____

Lesson 4: **Comparing and ordering angles**

• Estimate, compare, order and classify angles

1 Draw four different angles that are **greater** than a right angle.

2 Draw four different angles that are **smaller** than a right angle.

3 Order these angles, from smallest to greatest.

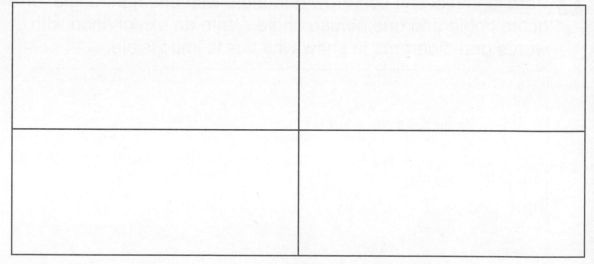

A   B   C   D   E   F

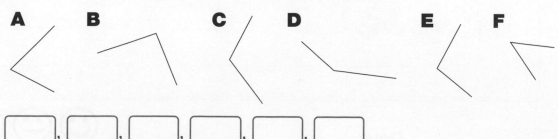

☐ , ☐ , ☐ , ☐ , ☐ , ☐

Geometry and Measure

4 Look at these shapes.

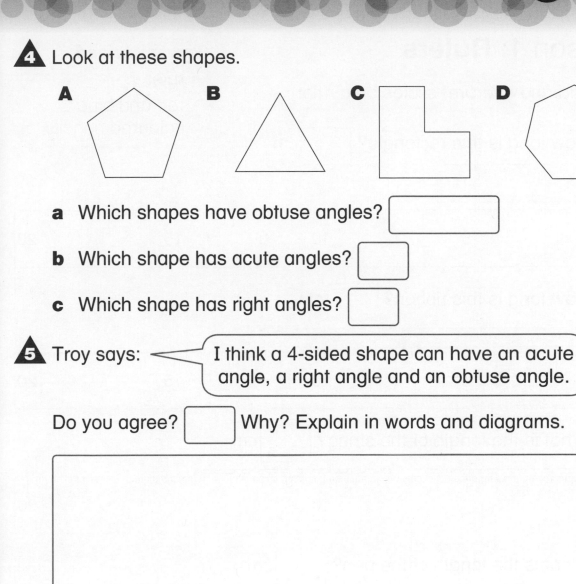

A   B   C   D

a Which shapes have obtuse angles?

b Which shape has acute angles?

c Which shape has right angles?

5 Troy says:

> I think a 4-sided shape can have an acute angle, a right angle and an obtuse angle.

Do you agree? Why? Explain in words and diagrams.

6 Draw an irregular heptagon. It must have two right angles, one obtuse angle and one acute angle. What are the other three angles?

Date: _____

173

Lesson 1: **Rulers**

Geometry and Measure

- Read and interpret scales on a ruler

1 How long is this rectangle? ⬚ cm

2 How long is this ribbon? ⬚ cm

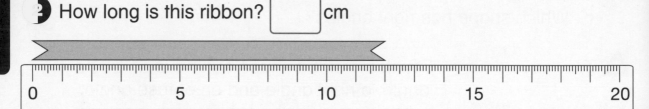

3 What is the length of the string? ⬚ cm

4 What is the length of the pen? ⬚ cm

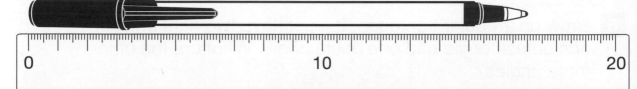

5 Use your ruler to draw lines of these lengths. Label your lines.

a $5\frac{1}{2}$ cm

b $9\frac{1}{2}$ cm

Geometry and Measure

c $12\frac{1}{2}$ cm

6 Ruby measures a length of string. She says that it is $\frac{3}{4}$ of a metre long. Stefan measures the same piece of string. He says it is 75 cm long. They are both correct. Explain why.

_____

_____

7 Find four objects to measure in the classroom. Estimate the length of each object then measure it. Don't forget to use the unit of measure: mm, cm or m.

Object	Estimate	Actual length

8 **a** Mark where the numbers 10, 20, 30, 40, 60, 70, 80 and 90 should go on this metre stick.

b Put a red mark where you think $\frac{1}{4}$ m is.

c Put a blue mark where you think $\frac{1}{2}$ m is.

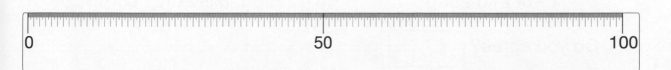

Date: _____

Lesson 2: **Finding mass**

- Use scales to find mass

1 **a** What is the mass shown on the scales? ____ g

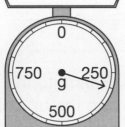

b What is the mass shown on the scales? ____ g

c Show 500 g on the scales.

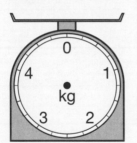

d Show $3\frac{1}{2}$ kg on the scales.

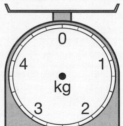

2 How many grams?

a 2 kg ____

b $2\frac{1}{2}$ ____

c $\frac{1}{2}$ kg ____

d $1\frac{1}{4}$ kg ____

e $\frac{3}{4}$ kg ____

f $2\frac{3}{4}$ kg ____

3 Barnie and Josie are finding the mass of potatoes.

Barnie says:

> My potatoes have a mass of 2 kg 250 g.

Josie says:

> Mine have a mass of 2250 g. Mine are heavier!

Do you agree? ____

Explain why.

_____

Geometry and Measure

4 a What is the mass of the toy car? ☐ g

b What is the mass of the teddy? ☐ g

c What is the mass of the apples? ☐ kg

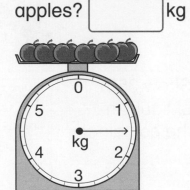

d What is the mass of the books? ☐ kg

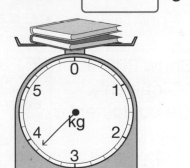

5 a Show 180 g on the scales.

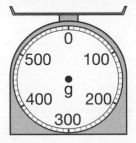

b Show $\frac{1}{4}$ kg on the scales.

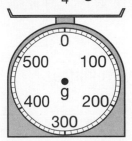

c Show $3\frac{1}{4}$ kg on the scales.

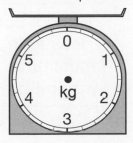

d Show $4\frac{3}{4}$ kg on the scales.

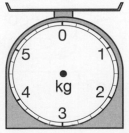

Date: _____

Lesson 3: **Measuring cylinders**

* Read and interpret scales on measuring cylinders

1 Write the amount of liquid shown in each container.

a [ ]

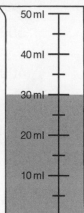

b [ ]

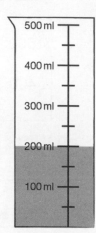

c Show 200 ml of liquid in the container.

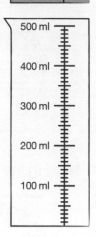

d Show 350 ml of liquid in the container.

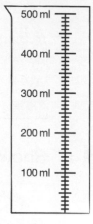

2 How many millilitres?

a $\frac{1}{4} l$ [ ]   **b** $1\frac{1}{4} l$ [ ]   **c** $2\frac{1}{2} l$ [ ]

3 Ruby and Luke are measuring the water in their bottles.

Ruby says: I've got $1\frac{3}{4} l$ of water.   Luke says: I've got 250 ml less than $2 l$. I must have more water than you.

Do you agree? [ ] Explain why.

Geometry and Measure

4 a Mark where you think 1 *l* 175 ml would be on this scale.

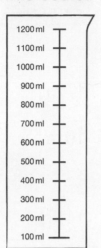

b How much is 2 *l* 175 ml in millilitres?

5 a Show 110 ml of liquid in the container.

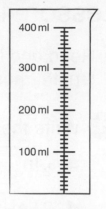

b Show 260 ml of liquid in the container.

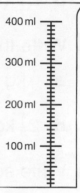

6 a Show 370 ml of liquid in the container.

b Explain how you know where 370 ml is on the container.

_____

_____

_____

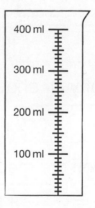

7 a What is the capacity of this measuring cylinder?

b How much liquid is in it?

c How much more is needed to reach the capacity of the measuring cylinder?

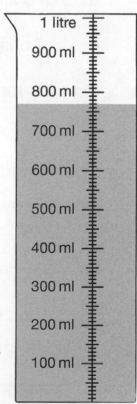

Date: _____

179

Lesson 4: **Revising measurement**

Geometry and Measure

• Read and interpret scales for length, mass and capacity

You will need
• scales

1 Write these lengths in centimetres.

a $\frac{1}{2}$ m [ ] cm   **b** $\frac{1}{4}$ m [ ] cm   **c** $1\frac{1}{2}$ m [ ] cm

d $2\frac{1}{4}$ m [ ] cm   **e** 1 m [ ] cm   **f** 10 m [ ] cm

2 Write these masses in grams.

a $\frac{1}{2}$ kg [ ] g   **b** $\frac{1}{4}$ kg [ ] g   **c** $1\frac{1}{2}$ kg [ ] g

d $2\frac{1}{4}$ kg [ ] g   **e** $\frac{3}{4}$ kg [ ] g   **f** $3\frac{3}{4}$ kg [ ] g

3 Write each capacity in millilitres.

a $\frac{1}{2}$ of a litre [ ] ml   **b** $\frac{1}{4}$ of a litre [ ] ml

c $\frac{1}{10}$ of a litre [ ] ml   **d** $\frac{3}{4}$ of a litre [ ] ml

4 Find the mass of your shoe using the scales. Record the mass:

a in grams [ ] g   **b** in kilograms and grams [ ]

5 Roshni and Laken are measuring the water in their bottles.

Roshni says: If I add another 250 ml of water to the 2*l* 750 ml in my jug, there will be 2*l* 1000 ml.

Do you agree? [ ] Explain why.

6 a Write the length of the pencil in millimetres [ ]

b Write the height of the fence in centimetres [ ]

c Write the length of the car in centimetres. [ ]

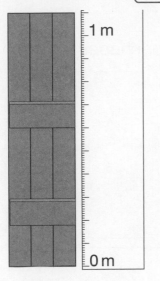

1 m

0 m

0 m   1 m   2 m   3 m

Geometry and Measure

7 **a** Draw a ring around the mass that is the odd one out.

$\frac{1}{2}$ of a kilogram   50 g   500 g   Why is it the odd one out?

_____

b Draw a ring around the capacity that is the odd one out.

1 l 250 ml   1250 ml $1\frac{1}{2}$ l Why is it the odd one out?

_____

c Draw a ring around the length that is the odd one out.

$2\frac{1}{4}$ km   2500 m   2 km 500 m   Why is it the odd one out?

_____

8 Oba says that $\frac{1}{2}$ a kilogram is lighter than 500 g.

4 Do you agree? [ ] Explain why.

_____

_____

Date: _____

Lesson 1: **Perimeter and area of 2D shapes**

Geometry and Measure

• Find the perimeter and area of 2D shapes

1 Write the perimeter of each of these regular shapes.

a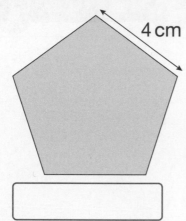
4 cm

b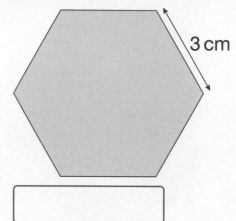
3 cm

2 Estimate the area of each shape.

a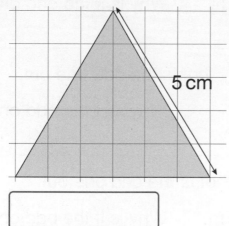
5 cm

b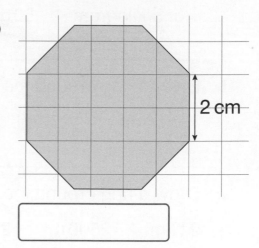
2 cm

3 Find the perimeter of a regular pentagon with sides of length:

a 7 m **b** 3 cm **c** 9 mm

4 Estimate the area of the circle.
Explain how you worked out your estimate.

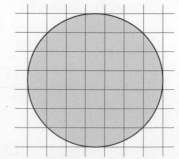

Geometry and Measure

5 Hattie and Eric each draw a regular shape. Hattie draws a 10-sided shape with sides of 4 cm. Eric draws a 9-sided shape with sides of 5 cm.

Hattie says:

I think that my shape has the longer perimeter because it has more sides.

Eric says:

The number of sides isn't important when working out perimeters. It's about the lengths of the sides.

a Who do you agree with? [ ] Why?

_____

_____

b What are the perimeters of their shapes?

Hattie [ ]   Eric [ ]

6 Maha draws a large regular octagon. It has a perimeter of 96 cm.

What is the length of each side? [ ]

Explain how you worked this out.

[ ]

7 Sara draws a large regular 9-sided shape. It has a perimeter of 144 cm. What is the length of each side? [ ]

Explain how you worked this out.

[ ]

Date: _____   ☺ ☹

183

Lesson 2: **Perimeter and area of rectangles and squares**

• Find the perimeter and area of rectangles and squares

1 Find the area of each rectangle.

a 3 cm

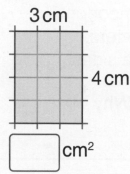

4 cm

☐ cm²

b 6 cm

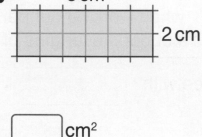

2 cm

☐ cm²

c 5 cm

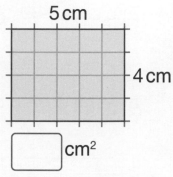

4 cm

☐ cm²

2 Find the area and perimeter of a rectangle with these dimensions:

a 8 cm by 4 cm

Area = ☐   Perimeter = ☐

b 9 cm by 7 cm

Area = ☐   Perimeter = ☐

c 5 cm by 6 cm

Area = ☐   Perimeter = ☐

d 12 cm by 3 cm

Area = ☐   Perimeter = ☐

3 Sasha draws a rectangle. It measures 12 cm by 3 cm.
Sketch it below then work out the area and perimeter.

☐

a Area = ☐   **b** Perimeter = ☐

Geometry and Measure

4 Tom draws a rectangle. It has a width of 4 cm and a length of 6 cm. He works out the perimeter using this calculation:
6 cm + 6 cm + 4 cm + 4 cm = 20 cm

a Explain how he could do this more quickly.

_____

He works out the area using this calculation:
6 cm + 6 cm + 6 cm + 6 cm = 24 cm^2

b Explain how he could do this more quickly.

_____

5 A square has an area of 144 cm^2. How long is each side?

6 A rectangle has an area of 48 cm^2. How long could each side be?

Give two possibilities.　　　　and

and

7 Thalia makes three squares all of the same size. They look like this.

Thalia thinks she can put these together to make other shapes of different areas. Is she correct? Explain your answer.

Geometry and Measure

Date: _____

185

Lesson 3: **Finding the area of 2D shapes (1)**

> • Find the area of compound shapes

Geometry and Measure

1 Work out an estimate for the area of each shape.

a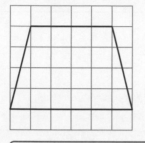

b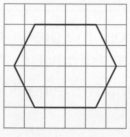

c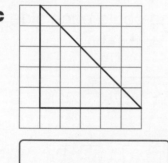

d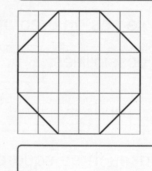

2 Draw a ring around the compound shapes.

a  **b**   **c**  **d** 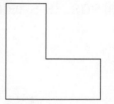 **e**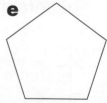

3 What is the area of each compound shape?
   ☐ = 1 square centimetre (cm²)

a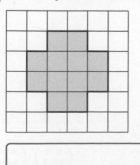

b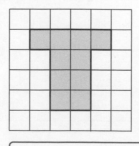

c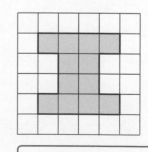

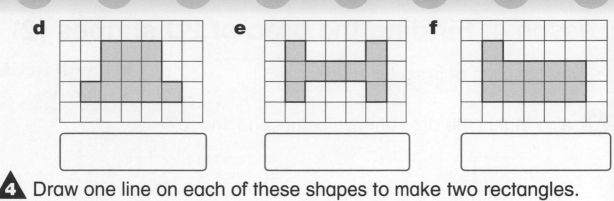

d · e · f

---

4 Draw one line on each of these shapes to make two rectangles.

5 Draw one line to show the possible rectangles you can make from this compound shape.

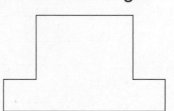

2 cm

8 cm

3 cm

10 cm

2 cm

5 cm

6 What is a compound shape?

_____

_____

_____

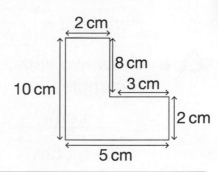

Sketch two different examples.

7 Sameer thinks:

4 Is this always, sometimes or never true? ☐

If the area of something gets bigger, so does the perimeter.

Explain your answer.

_____

_____

Geometry and Measure

Date: _____

☺ ☺ ☹

187

Lesson 4: **Finding the area of 2D shapes (2)**

- Find the area of irregular shapes

You will need
- ruler

1 **a** What is the area of each compound shape?

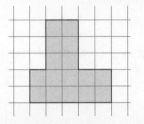 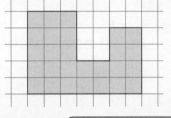

Area = [ ]   Area = [ ]

b Now work out the perimeter of each shape.

Perimeter = [ ]   Perimeter = [ ]

2 **a** Draw two vertical lines on this compound shape to make three rectangles.

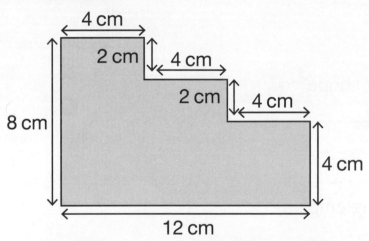

b Work out the area of each rectangle.

Area of 1st rectangle: [ ]

Area of 2nd rectangle: [ ]

Area of 3rd rectangle: [ ]

c Now work out the total area.

Total area: [ ]

Geometry and Measure

3 Lucy says:

> I think a compound shape is an irregular shape.

Jack says:

> I think a compound shape is a shape made from two or more other shapes.

Who do you agree with?

Why?

4 Samira says:

> I think that when we find the area of compound shapes made from rectangles we can use the generalisation length × width.

Is she correct?

Why?

5 Draw a compound shape in the box. You need to be able to divide it into three rectangles. Measure and label all the lengths to the nearest centimetre. Work out the area of your compound shape.

Area =

Now work out its perimeter.

Perimeter =

Date: _____

Lesson 1: **Compass points**

> • Use compass points to describe position, direction and movement

1 Fill in the missing compass points.

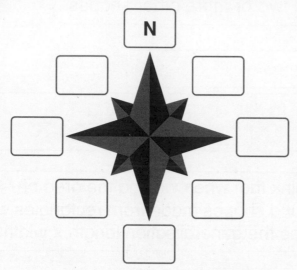

2 Write the eight compass points. Write the word and the letter or letters we use. The first one has been done for you.

north N, _____

_____

3 Use some of the words in **2** to write the shortest route through the white squares of the grid, from A to B. The route has been started for you.

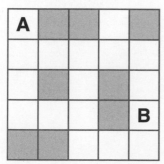

Move 2 spaces southeast. _____

_____

_____

Geometry and Measure

4 Sophie gives these instructions.

1. Move the counter 6 spaces SE.

2. Move the counter 4 spaces W.

3. Move the counter 3 spaces NE.

4. Move the counter 3 spaces N.

5. Move the counter 2 spaces E.

Colour the square where the counter finishes.

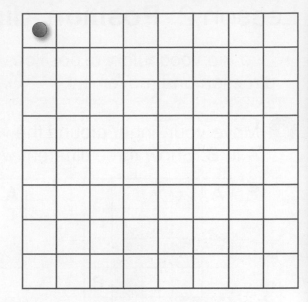

Geometry and Measure

5 Make up a set of five instructions to move the counter around the grid. Use cardinal and ordinal compass directions.

1. _____

2. _____

3. _____

4. _____

5. _____

Draw a line to show the path the counter takes. Colour the square where the counter finishes.

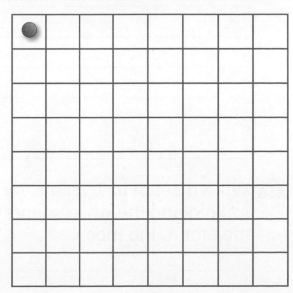

Date: _____

Lesson 2: **Position, direction and movement**

- Use the vocabulary of position, direction and movement

You will need
- coloured pencil

Geometry and Measure

1 Move your finger around the white squares of the grid to get from A to B. Show three different ways that you can do this.

a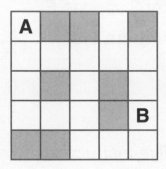

b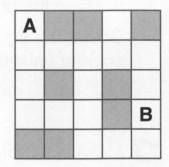

c

2 Write a set of instructions for a route along the white squares from the sun to the moon.

_____

_____

_____

_____

_____

_____

_____

3 Write a set of instructions for a route along the white squares from the star to the moon.

_____

_____

_____

_____

Geometry and Measure

4 Write a set of instructions for the **shortest route** along the white squares from the star to the sun.

_____

_____

_____

_____

5 **a** Shade five squares on each row of this grid. Create a path of white squares to move you from the triangle to the square.

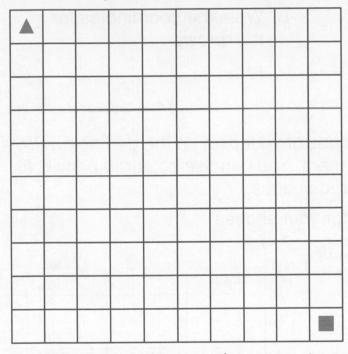

b Write instructions to show a route you can take along the white squares from the triangle to the square.

_____

_____

_____

_____

_____

Date: _____

Geometry and Measure

Lesson 3: **Coordinates**

- Use coordinates to describe position

You will need
- coloured pencil
- ruler

1 Look at this grid.

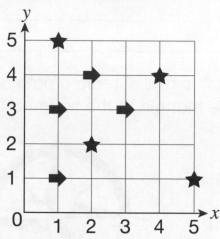

a Write the coordinates for the stars.

(____, ____) (____, ____)

(____, ____) (____, ____)

b Write the coordinates for the arrows.

(____, ____) (____, ____)

(____, ____) (____, ____)

2 **a** Draw eight small shapes of your own on the grid below. You decide where to put them. You can use coloured pencils to show different coloured shapes.

b List the coordinates for your shapes.

Shape	Coordinates

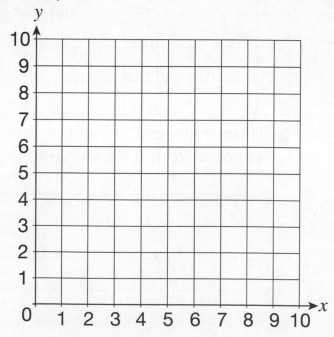

Geometry and Measure

3 **a** Write the coordinates for the circles.

(____, ____) (____, ____)

(____, ____) (____, ____)

(____, ____) (____, ____)

b Write the coordinates for the squares.

(____, ____) (____, ____)

(____, ____) (____, ____)

(____, ____) (____, ____)

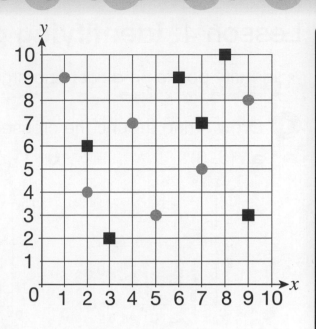

4 Plot dots at the points with these coordinates.

(8, 1), (9, 3), (7, 8), (3, 9), (1, 6), (2, 2), (5, 0)

The dots are corners of a shape.

Join them together.

What shape have you made?

_____

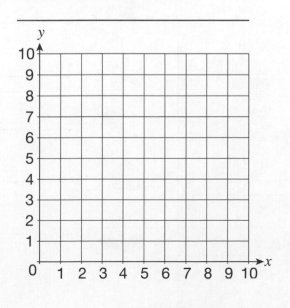

5 Think of a 2D shape. Plot the coordinates for the corners of your shape on the grid and list the coordinates.

_____

Join the coordinates.

What shape did you make?

_____

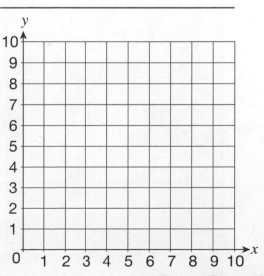

Date: _____

Geometry and Measure

Lesson 4: **Identifying points**

• Identify points on a coordinate grid

1 Draw a ring around the crosses that have been plotted **correctly**.

a (5, 3)

b (1, 2)

c (3, 4)

d (3, 4)

e (1, 2)

f (4, 2)

2 Write the coordinates of each object.

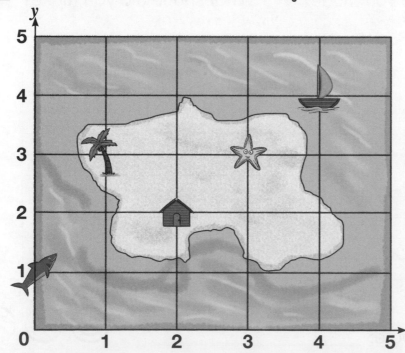

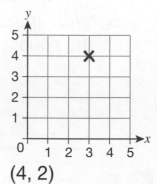 (___, ___)

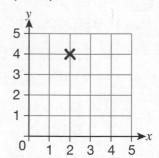 (___, ___)

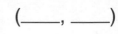 (___, ___)

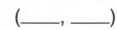 (___, ___)

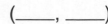 (___, ___)

3 Draw the object on the map at the coordinates given.

A shark at (1, 4)

A boat at (0, 3)

A hut at (3, 3)

A mountain at (3, 2)

A cave at (4, 2)

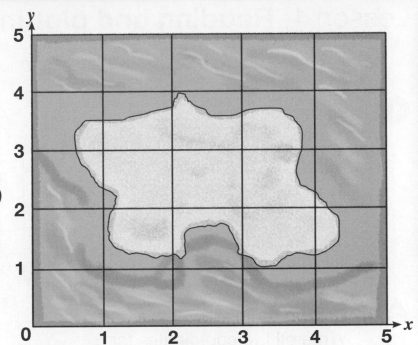

Geometry and Measure

4 Use the clues to draw the objects on the map.

An octopus at (1, 4)

A lake with an *x*-coordinate double that of the octopus and a *y*-coordinate half that of the octopus.

A boat at the same *x*-coordinate as that of the lake, but at a *y*-coordinate 3 greater than that of the lake.

A pair of mountains that are 2 squares apart on the *x*-axis and 1 square apart on the *y*-axis

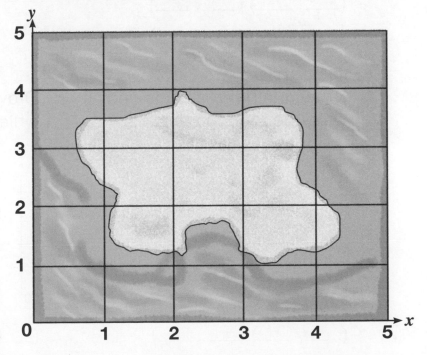

Date: _____

197

Lesson 1: **Reading and plotting coordinates**

Geometry and Measure

• Read and plot coordinates for vertices of shapes

1 What are the coordinates of the vertices of the square.

(____, ____) (____, ____)

(____, ____) (____, ____)

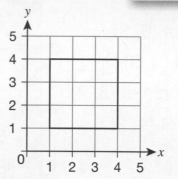

2 a Draw a rectangle on this grid. Write all the coordinates for the vertices of the rectangle.

(____, ____) (____, ____)

(____, ____) (____, ____)

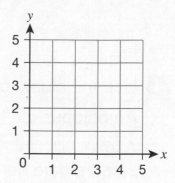

b Draw another rectangle and write the four coordinates.

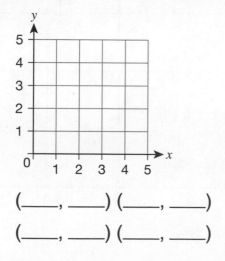

(____, ____) (____, ____)

(____, ____) (____, ____)

c Draw another rectangle and write the four coordinates.

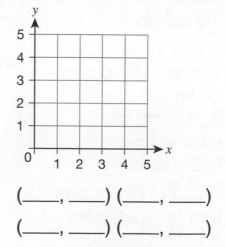

(____, ____) (____, ____)

(____, ____) (____, ____)

3 Draw an irregular pentagon on this grid. Each corner must be where two grid lines cross.

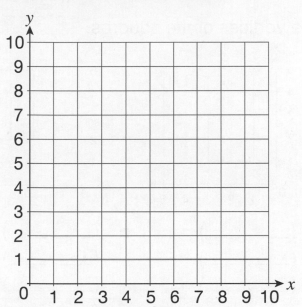

Write the five coordinates for the corners of your pentagon.

(____, ____) (____, ____)

(____, ____) (____, ____)

(____, ____)

4 Mandy says:

> If I plot the coordinates (1, 4), (3, 4), (3, 2) and (1, 2) I will be able to draw a square.

Do you agree? ☐

Use this grid to show your thinking.

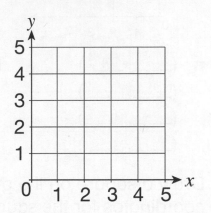

5 Draw these three shapes onto the grid and write the coordinates of the vertices of each shape.

a square

b irregular pentagon

c irregular hexagon

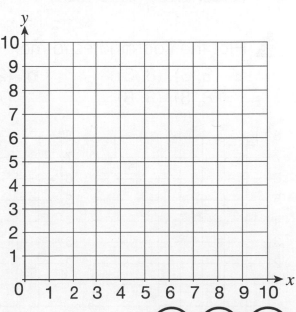

Lesson 2: **Shapes from coordinates**

• Read and plot coordinates for vertices of shapes

You will need
• ruler

1 Write the coordinates of the vertices of the squares.

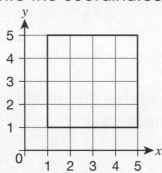 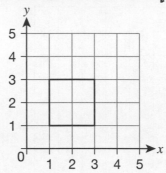 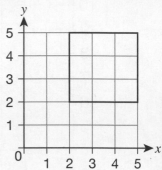

a (_____, _____) **b** (_____, _____) **c** (_____, _____)
  (_____, _____)   (_____, _____)   (_____, _____)
  (_____, _____)   (_____, _____)   (_____, _____)
  (_____, _____)   (_____, _____)   (_____, _____)

2 Draw a square on the grid. Write all the coordinates for the square.

(_____, _____) (_____, _____)

(_____, _____) (_____, _____)

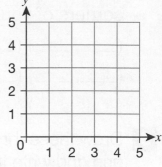

3 Plot the coordinates to make each shape.

a (3, 3), (8, 3),   **b** (2, 6), (7, 6),   **c** (1, 1), (1, 6),
  (8, 8), (3, 8)   (7, 3)   (8, 6), (8, 1)

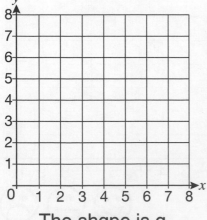 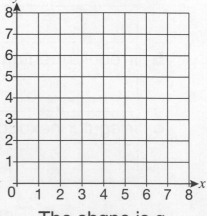

The shape is a

_____.

The shape is a

_____.

The shape is a

_____.

Geometry and Measure

4 Draw each shape on the grid. Write the coordinates of each vertex of the shape.

a Draw a right-angled triangle. Start at point (5, 2).

b Draw a square. Start at point (7, 7).

c Draw a rectangle. Start at point (0, 6).

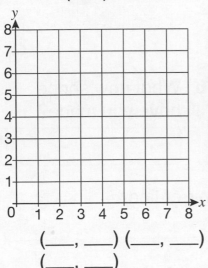

(___, ___) (___, ___)
(___, ___)

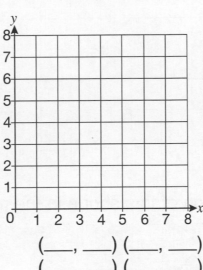

(___, ___) (___, ___)
(___, ___) (___, ___)

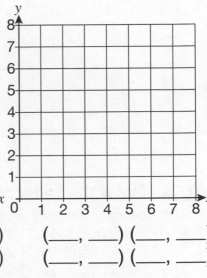

(___, ___) (___, ___)
(___, ___) (___, ___)

5 **a** Draw a rectangle. Write the coordinates of its vertices.

b Draw a triangle. Write the coordinates of its vertices.

c Draw a pentagon. Write the coordinates of its vertices.

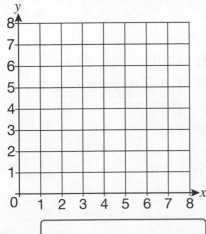

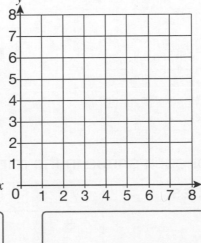

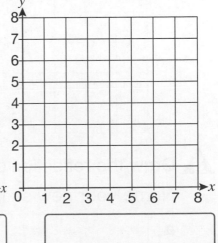

Date: _____

201

Geometry and Measure

Lesson 3: **Horizontal reflection**

• Reflect 2D shapes in a horizontal mirror line

You will need
• ruler

1 **a** Reflect this hexagon in the mirror line.

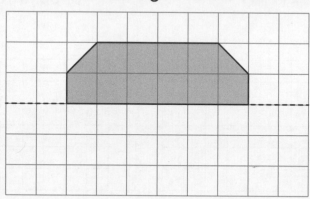

b What new shape have you made?

2 **a** Draw the reflection of this shape in the mirror line.

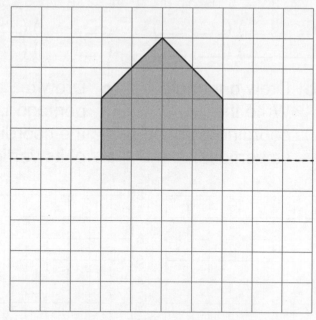

b What new shape have you made?

3 Draw the reflection of each shape in the mirror line.

Name your new shapes.

a

b

c

d

Geometry and Measure

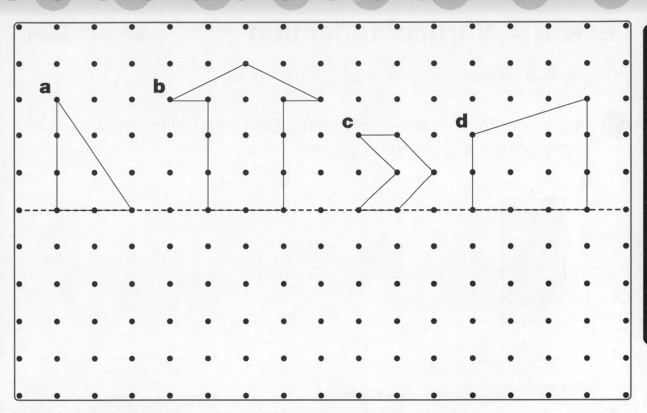

4 Reflect these irregular shapes in the mirror line.

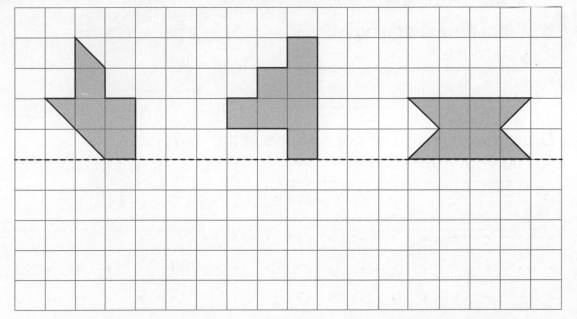

Date: _____

Lesson 4: **Vertical reflection**

Geometry and Measure

- Reflect 2D shapes in a vertical mirror line

You will need
- mirror
- ruler

1 Place your mirror on the mirror line. Draw the reflection. Write the names of the shapes you can see.

a

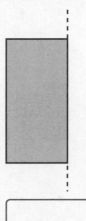

b

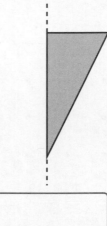

2 a Draw the reflection of this shape in the mirror line.

b What new shape have you made?

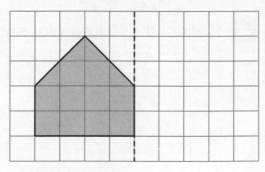

3 Draw the reflection of each shape in the mirror line. Name your new shapes.

a

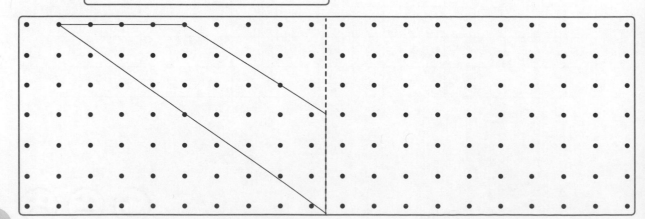

Geometry and Measure

b

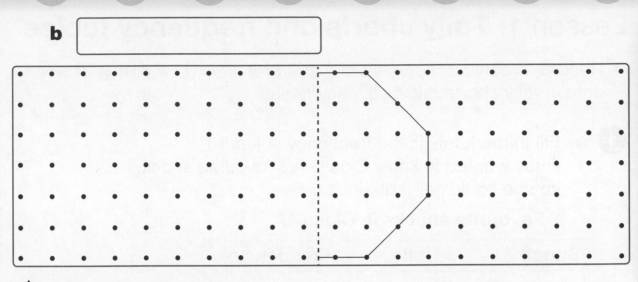

4 Draw the shape described to the left of the mirror line. Then reflect each shape.

Triangle with one right angle

a

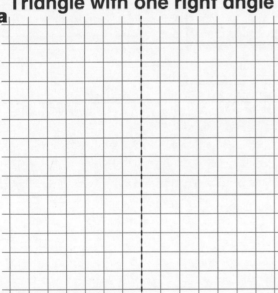

b **Rectangle**

c **Hexagon**

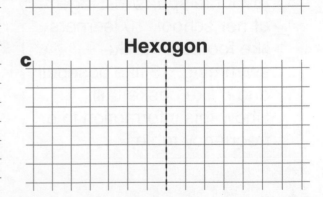

5 Reflect each of these irregular shapes in the mirror line.

a **b**

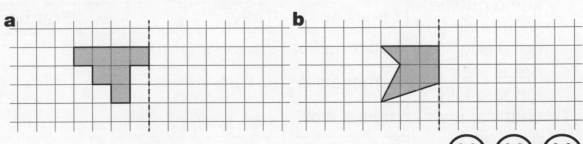

Date: _____

Lesson 1: **Tally charts and frequency tables**

Statistics and Probability

- Record, organise, represent and interpret data in tally charts and frequency tables

You will need
- ruler

1 a Fill in the totals in the frequency column.
Ruby wanted to know Class 4A's favourite snack, so she could sell it at lunch time.

Favourite snacks in Class 4A

Snack	Tally	Frequency
crackers	//	
popcorn	++++ ++++ ////	
fruit	++++ ////	
vegetables	++++	5

b Which snack should Ruby sell?

2 Tasha did a survey of favourite sports in Stage 4 at her school. 10 learners like football, 12 like swimming, 15 like baseball and 8 like basketball. Show this information in a frequency table.

3 Hassan asked some people: 'Where do you go to buy your bread?' He made a tally chart for their answers.

Buying bread

Location	Tally	Frequency
supermarket	++++ ++++ ++++ ++++ ///	
corner shop	++++ ++++ /	
market stall	++++ ++++ ++++	
bakery	++++ ++++ //	

a Count the tally marks. Write the totals in the Frequency column.

b Where do most people buy their bread? _____

c Which is the least popular location? _____

d How many more people buy bread at the supermarket than at the market stall? ☐

e How many people did Hassan ask altogether? ☐

4 Mr MacNeil collected the maths test results from 20 learners in Class 4. These were the results:
21, 27, 14, 5, 47, 35, 18, 5, 11, 34, 23, 32, 42, 45, 30, 48, 3, 45, 46, 32.
Organise these results in the grouped data frequency table.

Class 4 maths test results

Results	Frequency
0–9	
10–19	
20–29	
30–39	
40–49	

a The pass mark was 30. How many learners passed? ☐

b Those scoring less than 10 had to take the test again.
How many learners had to retake the test? ☐

c Explain how this frequency table works.

_____

_____

Date: _____

Statistics and Probability

Lesson 2: **Venn diagrams**

- Record, organise, represent and interpret data in Venn diagrams

You will need
- Survey results

1 Write the headings for this Venn diagram.

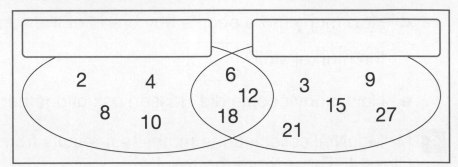

2 4 learners like swimming.
5 learners like running.
9 learners like doing both.

Write the numbers in the correct sections on the Venn diagram.

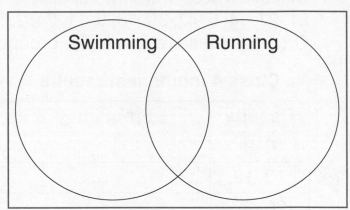

3 Write the numbers in the correct sections on the Venn diagram.

100   36   25

16   45   80

What does the overlap show?

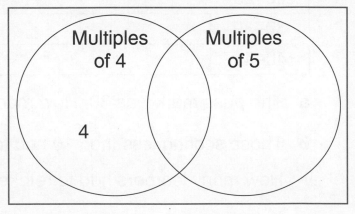

_____

_____

4 Mr Yip wanted to know what type of books learners prefer to read. He wanted to buy some to add to the school library. The categories were: fantasy fiction, realistic fiction and non-fiction.

a Use the Survey results to complete the Venn diagram.

Statistics and Probability

Class 4s favourite book types

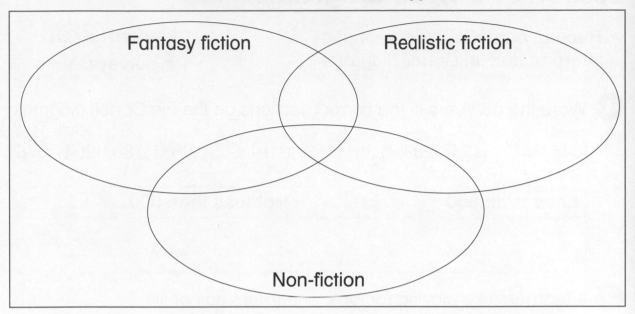

Fantasy fiction

Realistic fiction

Non-fiction

Use the data in the Venn diagram to answer these questions.

b What types of books does Pita like to read?

_____

c Which children like to read both realistic fiction and non-fiction?

_____

d What types of book should Mr Yip buy for the library?

_____

 5 **a** There are six mistakes in this Venn diagram. Draw a ring around them.

b Write down why they are mistakes.

_____

_____

Multiples of 3

Multiples of 5

66   18

54
32
72

81

27

24

16   36

Multiples of 8

_____

_____

Date: _____

Lesson 3: **Carroll diagrams**

• Record, organise, represent and interpret data in Carroll diagrams

You will need
• Survey results

1 Write the numbers in the correct sections on the the Carroll diagram.

345 417 124 208 736 453 319 602 863 780 204 576

Less than 500	Not less than 500

2 4 learners like playing football. 5 learners do not like playing football.

Write the numbers in the correct sections on this Carroll diagram.

Like football	Do not like football

How many learners were asked? ☐

3 Write the numbers in the correct sections on the Carroll diagram.

45   20   57   95   102   130   81   136

	Multiple of 5	Not multiple of 5
Even number		
Not even number		

What does the bottom right-hand section show us?

_____

4 Class 4 wanted to know what types of books learners prefer to read. The categories were: fantasy fiction, realistic fiction and non-fiction.

Statistics and Probability

a Use the Survey results to complete the Carroll diagram.

	Fantasy fiction	Not fantasy fiction
Non-fiction		
Not non-fiction		

b Write three statements from the data in the Carroll diagram.

_____

_____

_____

5 Imani made this Carroll diagram.

	Less than 20	Not less than 20
Odd	25, 11, 13, 15, 17, 19, 21, 23	10, 12, 14, 16, 18
Not odd		20

He has made mistakes.

a Write the numbers correctly in this Carroll diagram.

	Less than 20	Not less than 20
Odd		
Not odd		

b Write three statements from the data in the Carroll diagram.

_____

_____

Date: _____

Lesson 4: **Pictograms**

- Record, organise, represent and interpret data in pictograms

You will need
- paper
- ruler
- coloured pencils

1 Read the frequency table and key, then complete the pictogram.

Balloons sold

Colour	Frequency
red	4
blue	6
purple	8
pink	6
green	2

Balloons sold

Key
 = 2

red	blue	purple	pink	green

2 How many shirts of each colour does the school have?

Number of shirts

red = [   ]

blue = [   ]

green = [   ]

yellow = [   ]

Schools sports shirts

red	☺ ☺ ☺ ☺ ☺ ☺
blue	☺ ☺ ☺ ☺ ☺ ☺
green	☺ ☺ ☺ ☺
yellow	☺ ☺ ☺

Key: ☺ = 2 shirts

3 The members of the school Nature Club recorded how many butterflies they spotted in a day. Use the data in the frequency table to complete the pictogram.

Number of butterflies

Name	Frequency
Hiba	30
Youssef	45
Miriam	20
Joseph	10

Number of butterflies spotted in one day

Key
🦋 = 10

Hiba	
Youssef	
Miriam	
Joseph	

Statistics and Probability

Statistics and Probability

4 Class 4 planted sunflowers. They used a pictogram to record how many sunflowers they planted each day.

Sunflowers

Monday	
Tuesday	
Wednesday	
Thursday	
Friday	

Key: ✿ = 10 sunflowers

a How many sunflowers did they plant on Friday?

b On what day did they plant the most sunflowers?

_____

c How many more sunflowers did they plant on Wednesday than Thursday?

d Write one statement about the data in the pictogram.

_____

_____

5 A swimming club records the number of swimmers over a week.

Swimmers

Day	Number of swimmers
Monday	12
Wednesday	30
Friday	18

Use paper to draw two different pictograms. Choose a suitable symbol and decide on a different value for the symbol in each pictogram. Don't forget to include a title and a key with each of your pictograms.

Date: _____

☺ 😐 ☹

Statistics and Probability

Lesson 1: **Bar charts**

• Record, organise, represent and interpret data in bar charts

1 Use the data in the bar chart to answer the questions.

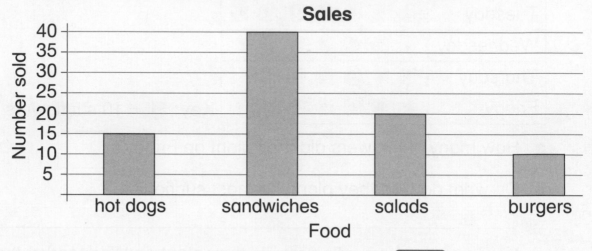

a How many sandwiches did the shop sell? [ ]

b How many more salads were bought than burgers? [ ]

c The owner bought 20 hot dogs to sell. How many did he have left? [ ]

2 Write three pieces of information that this bar chart shows.

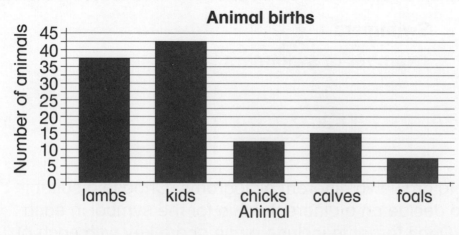

_____

_____

_____

3 Use the data in the pictogram to draw a bar chart. Use the key to decide on the intervals for the scale.

House builds

Month	Number of houses built
January	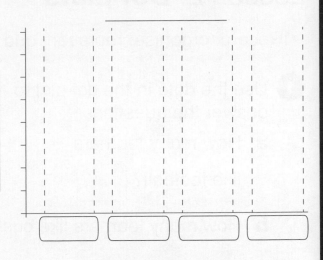
April	
July	
October	

Key: = 20 houses

4 Use the data in the frequency table to draw two bar charts, each with a different scale. Label one chart in intervals of 5 and the other in intervals of 10.

Favourite ice cream flavours of Stage 4 learners

Ice cream flavour	Frequency
chocolate	35
vanilla	40
strawberry	25

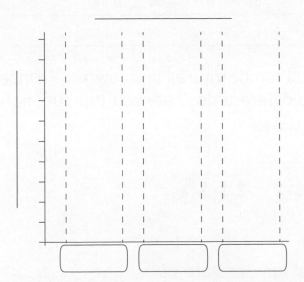

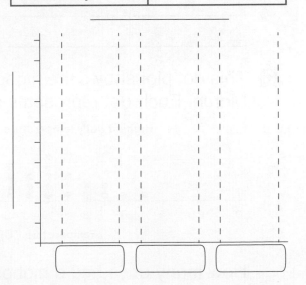

Mr Freeze wants to sell one ice cream in his shop. From the learners' responses' which should he choose?

Date: _____

Lesson 2: **Dot plots**

• Record, organise, represent and interpret data in dot plots

1 Use the data in the dot plot to answer the questions.

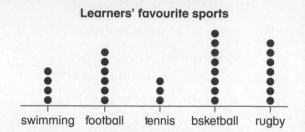

Learners' favourite sports

swimming football tennis bsketball rugby

a How many learners like football? ▢

b How many learners like basketball? ▢

c How many more learners like rugby than tennis? ▢

2 Write three pieces of information from the data on this dot plot.

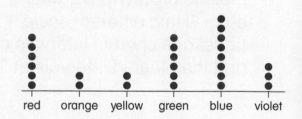

Learners' favourite colours

red orange yellow green blue violet

_____

_____

_____

_____

3 This dot plot shows the highest temperatures in a town in France in March. Each dot represents a different day. Answer the questions.

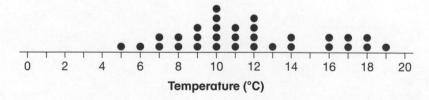

Highest daily temperatures in March

0 2 4 6 8 10 12 14 16 18 20

Temperature (°C)

How many days had a highest temperature of:

a 10°C? ▢   **b** 17°C? ▢   **c** Below 10°C? ▢

d At least 16°C? ▢   **e** Between 8°C and 13°C? ▢

f What was the highest temperature? ▢

4 Make up three of your own questions for the dot plot from **2**.

_____

_____

_____

_____

5 Use the data in the frequency table to draw a dot plot.

Learners' favourite subjects at school	Frequency
maths	8
English	4
music	5
art	7

maths   English   music   art

6 Use the data in the pictogram to draw a dot plot. Remember, each dot must represent one person.

People buying a tent at the camping shop this week	
Mon	♦
Tues	♦♦
Wed	♦♦♦♦♦
Thurs	♦♦♦♦♦♦
Fri	♦♦♦♦♦♦♦
Sat	♦♦♦♦♦♦♦♦
Sun	♦♦♦♦♦♦♦
Key: ♦ = 2 people	

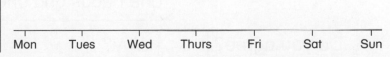

Mon   Tues   Wed   Thurs   Fri   Sat   Sun

Date: _____

Statistics and Probability

Lesson 3: **Chance (1)**

• Use the language of chance

1 Tick the statements that are **impossible**.

a In two years' time May will follow September. ☐

b It will rain next week in London. ☐

c I will go on holiday next year. ☐

d Next year there will be 33 days in October. ☐

e I will get up tomorrow morning. ☐

2 Write an example statement for each of these chance words.

Impossible _____

Maybe _____

Likely _____

Certain _____

3 Niall, Poppy and Tess are talking about tossing coins.

Niall says: If I toss a coin 20 times, I will get heads every time.

Do you agree? ☐ Why?

_____

_____

Poppy says: If I toss a coin twice, I am more likely to get one heads and one tails than anything else.

Do you agree? ☐ Why?

_____

Tess says: — What we get when we toss coins is all about chance. We don't know for sure!

Do you agree? ☐ Why?

_____

_____

4 Nellie says: — Chance is the likelihood of something happening. Certain means it will definitely happen.

Do you agree? ☐ Why?

_____

_____

5 Follow the instructions under each spinner.

a
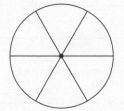

Write numbers on this spinner so that you have an even chance of spinning a 9.

b
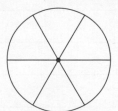

Write numbers on this spinner so that it is possible but not certain that you will spin a 2.

c
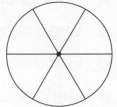

Write numbers on this spinner so that it is unlikely that you will spin a 4.

d Write numbers on these two spinners so that you have a better chance of spinning an even number on the first spinner than spinning an even number on the second spinner.

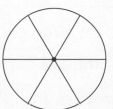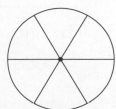

Date: _____

Statistics and Probability

Lesson 4: **Chance (2)**

You will need
• coin

Statistics and Probability

• Use the language of chance

1 Here are some digit cards:

| 6 | 7 | 8 | 9 | 9 | 9 |

Write good chance, poor chance or no chance beside each statement.

a I will pick a 3. _____   **b** I will pick a 9. _____

c I will pick a 6. _____   **d** I will pick a 7. _____

2 Write numbers in these cards so that it is certain you will pick an 8.

3 Write a statement for each of these probability terms.

a No chance _____

b Poor chance _____

c Even chance _____

d Good chance _____

e Certain _____

4 Look at the spinner. Tick the statements that you agree with.

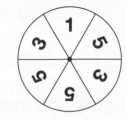

a I have an even chance of spinning a 5.

[ ] Why?

_____

b I have an even chance of spinning either an even number or an odd number. [ ] Why?

_____

c I have a poor chance of spinning a 1. [ ] Why?

_____

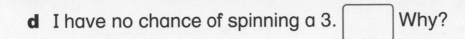
Statistics and Probability

d I have no chance of spinning a 3. [ ] Why?

_____

5 You need a coin. Toss the coin 10 times. Record your results.

[ ]

Were you surprised? [ ]

Why? _____

_____

_____

6 Two friends are rolling two 1–6 dice and adding the numbers
together to get a score.

Akachi says: < I have a good chance of scoring an even number.

Amare says: < I have a good chance of scoring an odd number.

Who do you agree with? [ ] Explain why.

_____

_____

7 Nola says: < I have this spinner. I think I have a good chance of spinning a 6.

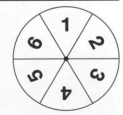

a Nola is not correct. Explain why?

_____

_____

b Write numbers on this spinner, so that Nola
will have a good chance of spinning a 6.

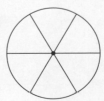

Date: _____

221

Acknowledgements

Photo acknowledgements

Every effort has been made to trace copyright holders. Any omission will be rectified at the first opportunity.

p118 Evikka/Shutterstock; p119 Evikka/Shutterstock; p156l Globe Turner/Shutterstock; p156c Artgraphixel/Shutterstock; p156r N. Vector Design/Shutterstock; p182 N. Vinoth Narasingam/Shutterstock; p220 Alexsandr Polle/Shutterstock.